THE
UNIVERSE

ANDREW COHEN

WITH A FOREWORD BY

PROFESSOR BRIAN COX

THE
UNIVERSE

WILLIAM
COLLINS

Previous page: The small open star cluster Pismis 24 lies in the core of the large emission nebula NGC 6357 in Scorpius, about 8,000 light years away from Earth. Captured by the Hubble Space Telescope, 2006.

William Collins
An imprint of HarperCollins*Publishers*
1 London Bridge Street, London SE1 9GF

WilliamCollinsBooks.com

HarperCollins*Publishers*
1st Floor, Watermarque Building, Ringsend Road
Dublin 4, Ireland

First published in Great Britain by William Collins in 2021

2024 2023 2022 2021
10 9 8 7 6 5 4 3 2 1

Text © Andrew Cohen 2021
Foreword text © Brian Cox 2021
Images © Individual copyright holders
Diagrams © HarperCollinsPublishers 2021

By arrangement with the BBC

The BBC logo is a trademark of the British Broadcasting Corporation and is used under licence

BBC logo © BBC 2014

Andrew Cohen asserts the moral right to be identified as the author of this work in accordance with the Copyright, Designs and Patents Act 1988

A catalogue record for this book is available from the British Library

ISBN 978-0-00-838932-1

Design: Zoë Bather
Editor: Helena Caldon

Printed and bound by GPS Group in Bosnia and Herzegovina

For my mum Barbara Cohen, it would be hard to find a kinder person in this or any other universe

ACKNOWLEDGEMENTS

When we embarked on the Universe television series in September 2019 we had, as always, a looming sense of the mountain in front of us, the familiar mixture of excitement and anxiety that always accompanies a project of this scale and ambition. What none of us knew back then was the size of the mountain that lay ahead.

By the time the first filming trip was ready to go in early 2020, the impact of COVID-19 and the impending pandemic was already massively affecting every aspect of our production and this, as we all now know, was just the beginning.

Over the last 18 months, every member of the Universe team has worked tirelessly to overcome the endless hurdles and challenges of making a landmark series during a pandemic. It's required huge dedication, skill and commitment to produce the series in circumstances that have made such huge demands on people's lives both professionally and personally. I am so grateful for everything the team has done to make such an exceptional piece of work in such exceptional times. As always this is just a small chance to say a big thank you.

First, thank you to Gideon Bradshaw and Jenny Scott who have led the team so brilliantly. As always Gideon's creativity and calm have endlessly shone through, helping create and craft a series with his editorial flare woven throughout. Whether it was flying out last minute to the Azores, crafting scripts, GFX and edits, or simply supporting the team in times of need – Gideon is in every sense of the word, world-class in what he does.

Equally, without Jenny and her amazing line management the series simply wouldn't exist. Navigating us through the endless challenges of COVID filming, she steered the series with huge expertise, resilience and kindness. Locations endlessly changing, myriad ever evolving quarantine rules and a new set of challenges for even the simplest location filming have been just a few of the mountains she has helped us climb. With Jenny and the team we are very lucky to have the best production management in the business, programmes simply don't happen without it.

We have also been incredibly lucky to have a world class team of film makers with Ashley Gething leading the way on two beautiful films, 'Stars' and 'Big Bang', Suzy Boyles delivering the brilliant 'Alien Worlds' and the initial script for 'Black Holes' before delivering her new baby boy Reuben, Kenny Scott showing us just how beautiful the UK can be to film in with his epic 'Islands of Light' and Tom Hewitson who stepped in to make the mind bending and beautiful 'Black Holes' film. We are very lucky to work with such talented people.

They were supported by a hugely talented team who have grappled with so many unprecedented challenges that have been overcome in the most creative of ways. So a very big thank you to Poppy Pinnock, Clem Cheetham, Milla Harrison, Sarah Houlton, Millie McNab, Emma Connor, Sophie Piggott, Chris Johnston, Paul Crosby, Graeme Dawson, Louise Salkow, Darren Jonusus, Nadine Limb, Martin West, Neil Harvey, Freddie Claire, Marie O' Donnell, Emma Ong and Vicky Edgar, and the many other people who have supported this production. Including a big thank you to Rob Harvey and all the team at Lola Post Production. And also to Nicola Cook and Nik Sopwith for the brilliant development work that allowed us to make The Universe.

A very special thanks to Laura Davey, who worked tirelessly in leading this and every production in the Science Unit over the last 18 months, building the foundations upon which all of our creativity and ambition has been able to flourish.

Finally, a big thank you to the team at William Collins. Once again you have produced the most beautiful book and added so much more value to simple words. So thank you to Helena Caldon, Hazel Eriksson, Zoë Bather and Chris Wright. And of course, a very big thank you to Myles Archibald – apart from a quick visit to the emergency dentist, it's been a painless pleasure.

FOREWORD

Contemplating the Universe is exhilarating and terrifying in equal measure. Nobody, professional scientist or otherwise, understands how to internalise the fact that we exist – alone, as far as we can tell – in a void illuminated by 2 trillion galaxies. Perhaps that's why, as I say in my introduction to the television series on which this book is based, there is a sense of relief that rises with the dawn. The brightening sky hides the stars and the questions they pose. What questions? Questions that run the risk of sounding naive because they are profound. Questions, perhaps, that we feel we should not ask in educated company.

Wonder is unfashionable amongst fashionable cynics. How did the Universe begin? How will it end? What is the meaning of it all? Are these questions childlike? Maybe. Childish? Certainly not. The first two questions are eminently askable because they fall comfortably within the domain of science. That's not to say we are able to answer them, but the study of the origin and fate of the Universe is the province of cosmology, the foundation of which is Einstein's Theory of General Relativity, published in 1915. For 100 years we have possessed a scientific theory that forces us to contemplate the beginning of time, supported by a century of astronomical observations. We live, certainly, in an expanding universe. The distances between the galaxies are increasing today, and that implies that in the past the galaxies were closer together. If we run the equations backwards, we find that everything was very close together 13.8 billion years ago, and we call that moment the Big Bang. We have even detected the afterglow of the Big Bang as ancient light known as the Cosmic Microwave Background Radiation. In the old days, I'd tell you that it formed part of the static fuzz on a detuned TV set and you could have watched the ghosts of creation dance for yourself.

We know, therefore, that the Universe has not always been like the one we observe today in our cosmic neighbourhood: old, cold, almost empty, save for occasional islands of light scattered as snowflakes on the wind. Long ago, it was hot and dense and highly energetic, and out of those fires we emerged.

Opposite: The rocket carrying NASA's Mars 2020 Perseverance rover launched in July 2020, ushering in a new era of exploration.

The story of our emergence is written today not in ancient texts but in textbooks, because we have observed it rather than invented it. The way to understand nature is to look at it. Inside our supercomputers we build simulated universes and watch as great lanes of dark matter condense as morning dew on a spider's web in patterns laid down by sub-atomic-scale fluctuations when our universe was far less than a billionth of a second old. We see stars and galaxies form around the scaffolding of the web and planets condense from the leftovers. On our planet we have seen active geology encourage carbon atoms to form into long-chain molecules that encode information, and we understand in broad scope how evolution by natural selection allowed those molecules to write symphonies.

This series was filmed in unprecedented times, during a global pandemic predicted by science and ultimately alleviated by science. I think the experience shaped the series and made it more philosophical and perhaps more polemical. After all, the reliable knowledge upon which our understanding of genetics and viruses and vaccines rests was not acquired by fashionable cynics but by people driven by wonder. The foundations of that wonder lie amongst the stars because astronomy is the oldest science, and therefore a heartfelt celebration of astronomy feels both apposite and necessary. Today, the questions posed by the heavens still stretch the human mind beyond its capabilities, and without such challenges to our insular vanity there will be no progress. From the greatest of galaxies to the supermassive black holes at their heart, the sky is populated by natural objects we don't yet understand, and that's why the sky is so valuable a resource. Who knows where the new knowledge will take us, but long experience tells us it will lead to something wonderful.

I hope you enjoy the series and the book and, perhaps, are motivated to look upwards into the night not in terror at the infinite spaces but in exhilaration at the infinite wonder. What is the meaning of it all? I don't know, but we won't find the answer by looking inwards. We'll find it by lifting our gaze above the horizon and outwards into the Universe beyond. We used to look to the sky and see only questions. Now, we are beginning to see answers.

Professor Brian Cox, 2021

Next page: This giant red nebula (NGC 2014) and its smaller blue neighbour (NGC 2020) are part of a vast star-forming region in the Large Magellanic Cloud. Captured by the Hubble Space Telescope, 2020.

Opposite: The International Space Station is an example of the very best of humanity; collaboration in the name of progress and discovery.

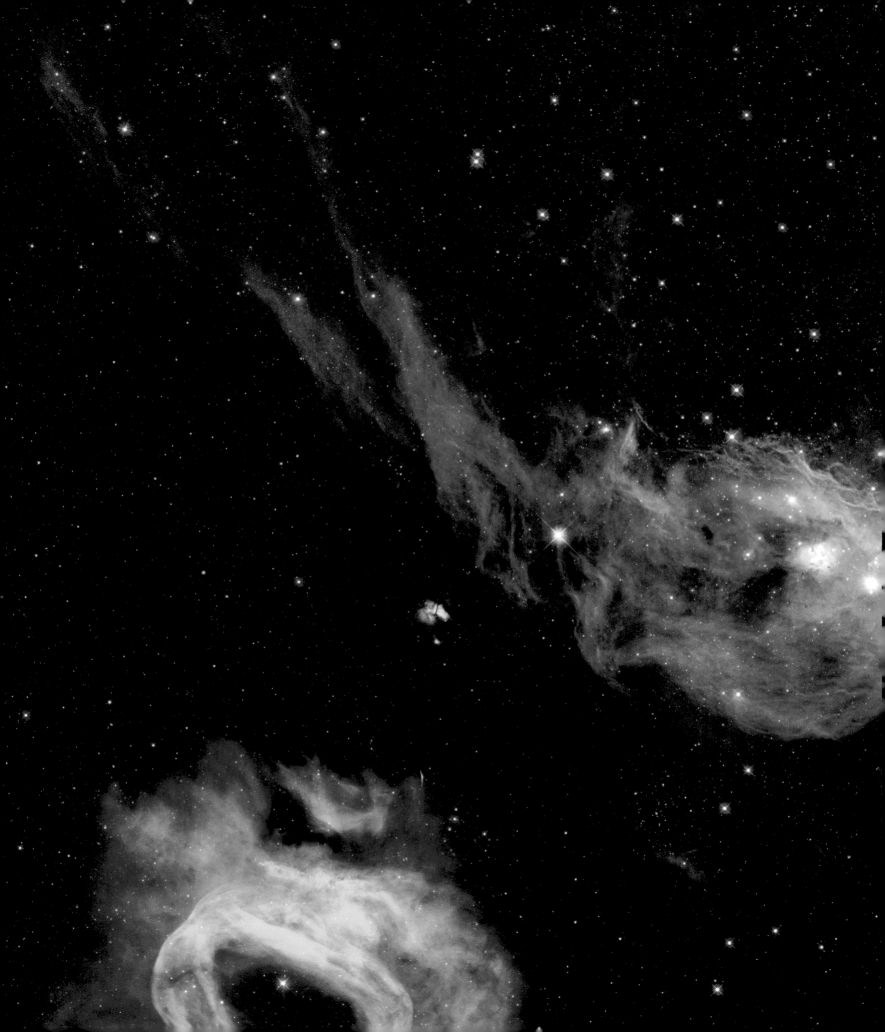

ALIEN WORLDS

'Where is everyone?'
Enrico Fermi

THE LONELY WALK HOME

There is no grander question than 'are we alone?' The implications of this are immense and endlessly unnerving and perhaps summarised best by the quote attributed to Arthur C. Clarke: 'Two possibilities exist: either we are alone in the Universe or we are not. Both are equally terrifying.'

In the last 40 years we have made the first tentative steps to answering this terrifying question. Yet the answer will not lie in great galactic structures, or among the bright and massive stars that light the night sky; instead it'll be found on the tiniest of objects in the Universe. Things that together make up less than 1 per cent of the Universe's mass: planets.

We now know our own galaxy and almost certainly every galaxy in the Universe is full of alien worlds. The discovery of these worlds has not yet led us to be on the verge of meeting another advanced civilisation, but it has allowed us to explore worlds with an endless variety of characteristics and so paint the possibilities of life onto an infinite array of settings. Planets are the Universe's chemistry sets. A place where the elements of the cosmos can come together, all squished up against each other in the right state and with a ready supply of energy, poised to make something new, to trigger the cascade of chemical reactions that ultimately make and run life. Although in cosmic terms planets are tiny or insignificant (as some of the smallest objects in the Universe), they are also unique and special – they are the only place where meaning can arise.

Across the history of human civilisation, we have spent a surprising amount of time believing that we share the Universe with other beings and not just existing in an infinite expanse of isolation. Around 2,500 years ago, Greek philosophers such as Democritus and Epicurus looked up into the night sky and imagined an infinite

Above: Ancient philosophers including Democritus (top) and Fakhr al-Din al-Razi (his work above) imagined a universe of infinite worlds.

Above: Renaissance astronomer Nicolaus Copernicus broke away from the idea of the Earth being at the centre of the Universe, proposing the Sun in its place.

universe populated with an infinite number of worlds. It was a view that would be shared by many, including a long line of medieval Muslim scholars such as Fakhr al-Din al-Razi, who argued for the existence of a universe filled with 'a thousand, thousand worlds beyond this world'.

But ultimately this expansive outlook was overridden by the Earth-centric view of the Universe that came to dominate at least Western thought for well over 1,500 years. It wasn't until Copernicus broke out of our bubble of self-importance and opened our eyes to a cosmos that wasn't simply constructed around us that we began to accept the inevitable consequence of an infinite universe. Slowly but surely, from the middle of the last millennium, great thinkers began to share in private and occasionally in public a belief in the existence of worlds beyond our own. Such thoughts were not without significant dangers, with the scientific revolution never bubbling far away from accusations of heresy. Perhaps most famously, Giordano Bruno, the Italian friar and cosmologist, was burnt at the stake with his tongue pinioned – in part to prevent him from repeating his heretical belief that the Universe is infinite and filled with an infinite number of habitable worlds.

Slowly, as astronomy advanced our knowledge of the Universe, the 'ever-receding horizon' brought with it an expanding acceptance of our place in a universe filled with planets and, potentially, life. As the age of enlightenment filled our minds and imaginations with the knowledge of a universe bursting with wonders, the plurality of worlds became a commonplace belief amongst the intellectual classes of Europe. William Herschel, Benjamin Franklin and Camille Flammarion are just a few of the esteemed names from the eighteenth and nineteenth centuries who argued for a multi-world universe. (Flammarion was particularly influential in the arguments

'Innumerable celestial bodies, stars, globes, suns and Earths may be sensibly perceived therein by us and an infinite number of them may be inferred by our own reason.'
Giordano Bruno, De l'infinito, Universo e Mondi *(1584)*

Opposite: Medieval illustration of the Ptolemaic system known as the geocentric model, which suggested that the Earth is the centre of the Universe.

Right: Woodcut from Giordano Bruno's last work, *On the Composition of Images, Signs and Ideas* (1591) in which he put forward some of his 'heretical' ideas.

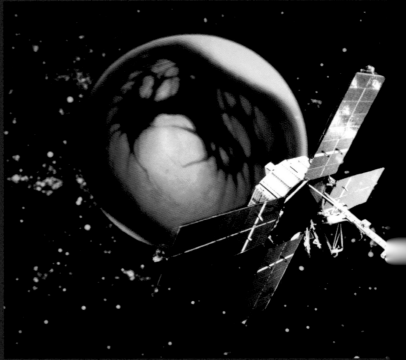

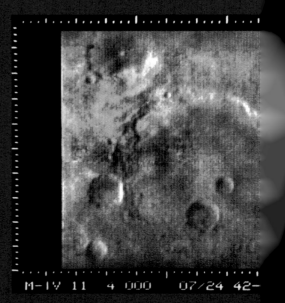

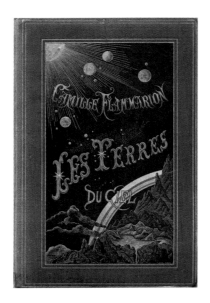

Above: An example of Camille Flammarion's otherworldly woodcuts, 1884.

'It may just be that life as we know it – with its humanity – is more unique than many have thought.'
President Lyndon Johnson

he made, both as a scientist and as an author of some of the earliest science-fiction stories – stories that painted vivid pictures of exotic worlds with sentient planets and aliens.)

It seemed the more we looked up into the skies above us, the more evidence we found to support the existence of our cosmic neighbours. So much so that by the beginning of the twentieth century we gazed at our nearest neighbouring planet, Mars, with a collective belief that there were canal systems built across the red planet by a Martian civilisation. Sometimes seeing more can actually make you see less, and although the telescopes we trained on the red planet became increasingly powerful, our interpretation of the images they provided began to drift further towards hope than reason. This would all change, though, with the direct exploration of our own back yard that began through the latter half of the twentieth century. For the first time in human history we would not be just looking up and wondering about the alien worlds above us, we would actually be seeing in close-up detail the surfaces of planets such as Mars and Venus.

With the launch of the Mariner 4 spacecraft on 28 November 1964, we were at last bound for Mars, with the overriding expectation that we would discover a planet not too unlike our own. After eight months of navigating by the stars and travelling further than any human object in history, Mariner finally reached its destination. Powering up its cameras to record the first ever images of another planet, back on Earth we waited for those precious images to emerge.

But as the raw data beamed back across more than 200 million kilometres of space and slowly transformed into that first image of the Martian landscape, hope of finding a habitable world faded.

Mariner 4 looked down onto a planet without a drop of water on its surface, and with little or no atmosphere, revealing a barren, cold and dead world. Life on Mars had been the subject of huge speculation and science-fiction stories for decades, yet Mariner 4 had found none. Our first planetary visit had shattered our understanding of what a planet could be and with it the notion of our own place in the Universe.

Over the last few decades we have explored our own solar system in ever more detail, with the discovery of life at the forefront of many of these missions. We have visited all of the planets and many of the moons that held some glimmer of biological hope, but over and over again our robotic explorers have found worlds devoid of the life we dreamt of finding. There are no advanced civilisations, animals or plants on any of our neighbouring worlds. Instead, our only remaining hope is that they may host life in its simplest forms, tiny single-celled bacteria, hidden in the warmth of oceans or beneath an icy landscape.

For much of the second half of the twentieth century we looked out into a void, the myopia of our scientific knowledge convincing us that we were once again alone. With no ability to look beyond the seven other planets and multiple moons of our own solar system, we mixed up the unknown with the unknowing. The bright glare of the stars blinded us from seeing what else might be out there, a darkness shrouded with mystery. But since then a revolution has occurred; a journey of discovery that has exposed the darkness and transformed our understanding of what lies in the shadows. From an empty void to a cosmos crowded with worlds, each is a new place in which to look for alien life.

A NEW WORLD ORDER

ANATOMY OF A PULSAR

Neutron stars, with their powerful magnetic fields, can rotate very quickly, accelerating charged particles by the magnetic field and releasing radiation from the poles. This sends two photon beams across the sky, detected as pulses on Earth.

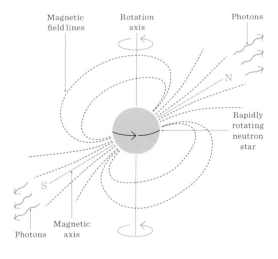

For humankind, the Universe became a very different place on 9 January 1992. For decades astronomers had been developing techniques to detect a planet orbiting a star outside of our own solar system, but despite a smattering of claims that emerged from the 1950s onwards, none of them stood up to scrutiny. We had assumed there must be planets out there, but we had no way of proving the hunch. But that was all about to change with a discovery that was made around one of the strangest types of star in the night sky.

Neutron stars are formed when a giant star explodes in a supernova but then collapses back in on itself. If the star is big enough a black hole can form, but sometimes the core doesn't totally collapse and instead it creates an incredibly dense object known as a neutron star. One such neutron star is PSR B1257+12, also known by the name Lich (an Old English word, meaning corpse), which was discovered in 1990 using the Arecibo Telescope. This tiny object has a radius of just 10 kilometres but a mass that is equivalent to almost one and a half times that of the Sun and a scorching surface temperature of 28,582 degrees Celsius. But that's not the end of the strangeness of this incredible object; PSR B1257+12 is a particular type of neutron star called a pulsar, a highly magnetised type of neutron star that spins at an incredible rate (9,659 rotations a minute in the case of PSR B1257+12) and in so doing emits a pulsating beam of electromagnetic radiation from out of its poles. Here on Earth that means we can observe these incredibly powerful and predictable pulses and precisely measure the interval between them.

Since they were first discovered by Jocelyn Bell Burnell and Antony Hewish in 1967, these galactic clocks have become extremely useful objects in helping astronomers explore a wide range of theories, from the interstellar medium to gravity waves, all by measuring disturbances to the rhythmic pulse.

Back in the early nineties, as astronomers Aleksander Wolszczan and Dale Frail monitored the newly discovered pulsar PSR B1257+12, they noticed its pulse was occasionally a little off. It should have been emitting its beacon of light every 0.006219 seconds but instead it seemed to be disrupted by something and that something was not entirely random. It was an oddity that had never been seen, or heard, among any other pulsar in our galaxy. So the data was double-checked, then triple-checked, but the anomaly remained.

The off beats came at regular intervals, suggesting some other predictable element was interfering with the signal. There remained only one plausible explanation: something was dragging the pulsar back and forth, exactly as the Earth drags on the Sun. The effect of this drag was causing an irregularity in how fast its pulses of radiation reached us.

Opposite: The Arecibo Telescope, in a natural crater in Puerto Rico, examined planets, asteroids and the Earth's upper atmosphere.

Left: One of the first images of an exoplanet (red) orbiting a brown dwarf star (white). The star, called 2M1207, is around 170 light years away.

'If you lived on a planet orbiting a pulsar, you would want to make sure to be very far away. Although these stars are very small – they can be as small as a city – they're extremely dense and have an enormously powerful gravitational field. Pulsars are so dense that a teaspoon of their neutron material would weigh as much as a mountain here on Earth.'
Nia Imara, Astrophysicist, University of California

On 9 January 1992, after months of work, Wolszczan and Frail had an explanation for the odd behaviour of this tiny stellar remnant flashing in the darkness 2,300 light years from Earth – two tiny planets were orbiting around it every 67 and 98 days respectively, exerting the tiniest of gravitational tugs that could be seen in the irregular heartbeat of the pulsar. Poltergeist and Phobetor, as they came to be known, were the first planets discovered outside our solar system – the first indication of a universe of worlds.

Two years later, another planet, Draugr, was detected and added to the system, a planet less than half the mass of our moon that was orbiting with its larger siblings. But for all the excitement, the discovery of these three first exoplanets changed nothing in terms of our lonely view of the night sky. Orbiting around the intense radiation of the Lich pulsar, Draugr, Poltergeist and Phobetor are planets, but not in a form that we would recognise from the hospitable realm of our own solar system. Created from the recycled dust of a previous generation of worlds that would have circled the star before it exploded, these planets are trapped in a star system that is more violent and destructive than we can imagine. Poltergeist, which is four times the mass of the Earth, orbits its star every 66 days and is consumed by radiation that may even charge whatever atmosphere it has, painting the sky with the most beautiful of auroras. But this is no Eden; its cold, dead surface is battered by intense radiation, just like its rocky sibling, Phobetor. No liquid water can exist on these worlds, and no life could ever survive here.

These first exoplanets were remarkable objects, chance discoveries made by following the strangest of signals that led to a profound shift in the way we understood the Universe, and in an instant our tally of planets that we know of in the Universe went from eight to 10.

But there was more to this discovery than just numbers. The very nature of Poltergeist and Phobetor told us something profound about the Universe. These were planets that had not formed around a newly born star like our own, but around a dying star created from the ghostly remnants of its past life. Their unlikely story suggested a fundamental truth about the formation of planets in the Universe – anywhere that there is enough material, enough energy and enough gravity, planets will be born. The Universe is virile. Our vast 1,000,000,000,000,000,000-kilometre-wide Milky Way must harbour hundreds, thousands, maybe even millions of planets born some time in its 13-and-a-half-billion-year history. Hidden in the darkness, just waiting to be found. With this knowledge the race to find the next exoplanets had begun.

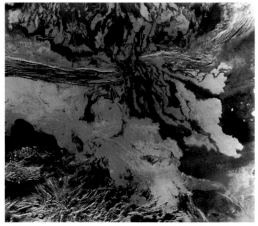

Left: The uninhabitable surfaces of Mercury (top) and Venus (bottom). Both planets have been ravaged by their proximity to the Sun.

Sara Seager, Professor of Planetary Science, Physics, Aeronautics and Astronautics, MIT

We've been asking questions for millennia. We know there are other stars, other suns. We know the stars are part of a galaxy. We know our galaxy is one of hundreds of billions of galaxies out in our universe. But we still don't know everything. We don't know about all the planets around all those stars, how they formed. We don't yet know if there are other solar systems out there like ours and we don't know if there are other Earths with life on them.

We humans are born explorers. We want to find other planets because we want to know if there's any life out there. And planets, rocky planets like Earth, are the place to search. The ingredients for life appear to be everywhere, just floating around in outer space. But these ingredients can only concentrate on a planet, and we need the planet to concentrate molecules and energy and everything to make life start and happen.

We're doing our best to search for life, but right now and for the foreseeable future, we can only search around the very nearest stars to us – so only a dozen or 100, or at the very most a thousand stars, and there are hundreds of billions of stars in our galaxy. It would be like being able to meet all the neighbours on your block. You could never meet anyone from another city. So if there is life out there, we may not be able to find it.

The search for exoplanets really got going in the mid-1990s, when astronomers found the first exoplanets orbiting sun-like stars. But these planets are like nothing in our own solar system. They're so strange that at first people didn't want to believe they even existed. Astronomers found giant planets very, very close to the star, nearly ten times closer to their star than Mercury is to our sun. And a planet has no business being there. There's not enough material around a forming star to make a planet that close in.

As more and more planets were discovered, it was harder and harder to call them something other than a planet. We didn't see the planets directly. We only saw the planets' gravitational effect on the star, because as the planet and star can orbit the common centre of mass, you can think of it like the planet tugging on the star. We can measure the stars' line of sight motion. We can measure the stars' wobble due to the orbiting planet. Now, later on, there came a very, very special event. If a planet star system is aligned just so, just perfectly so that the planet orbits in the plane, from our viewpoint, the planet might go in front of the star or transit as seen from our telescopes, and the transit can cause a tiny drop in brightness of the star.

So as we discovered more and more planets, the chance that one of them would have this very special alignment to show transits was increasing. And finally, a planet did transit and was observed. And there's nothing else that could indicate. There's nothing else that could cause a wobble and a transit of the same star that matched up perfectly. So after that, there was no doubt whatsoever.

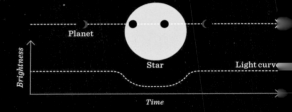

As a planet crosses in front of its parent star (seen from Earth here), transit photometry will analyse the dip in the star's light curve, which can identify a planet in orbit, and gauge its size as well as the composition of its atmosphere.

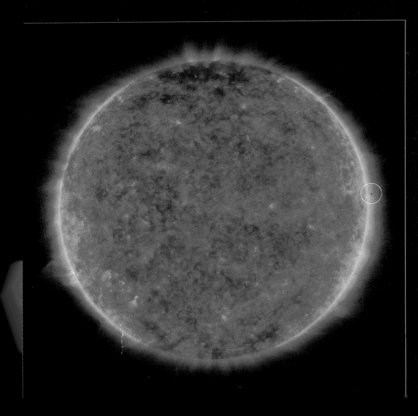

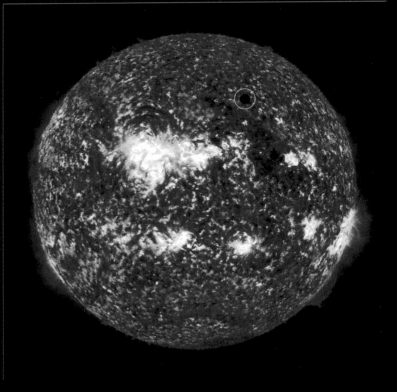

THE RACE FOR NEW WORLDS

Poltergeist and Phobetor radically changed our perspective of the Universe, but perhaps not our sense of loneliness. To do that we needed to find a planetary system a little more like our own, planets that we recognised to be in some way similar to the eight within our own solar system, orbiting around a main-sequence star like the Sun. Perhaps even a rocky planet with the potential for liquid water to pool on its surface.

Astronomers across the world raced to train their telescopes on sun-like stars, desperate to find an alien world that could be Earth's kin. To look beyond the glare of living stars and into their shadows to see the planets that we knew must be out there. Something that would help us understand whether our world was unique or the norm, whether we were significant or insignificant.

Around 50 light years away from Earth, 51 Pegasi is an unremarkable main-sequence star in the constellation of Pegasus. At least 2 or 3 billion years older than the Sun, this yellow G-type star is heading towards the end of the hydrogen-burning period of its life on its journey to the next stage as a red giant.

In January 1995, a Swiss PhD student called Didier Queloz sat 500 trillion kilometres away from 51 Pegasi, at the Haute-Provence Observatory in the south-east of France. Queloz was working with his supervisor, Michel Mayor,

Above: Michel Mayor, who discovered 51 Pegasi b, the first confirmed exoplanet, pictured at the Astrobiology Centre, Madrid.

Below: The Haute-Provence Observatory, where Didier Queloz and Michel Mayor first spotted 51 Pegasi's tell-tale wobble.

Opposite: Artist's impression of a hot Jupiter-class exoplanet, encircled by clouds.

on a newly developed planet-hunting system called ELODIE, recently installed at the observatory. Designed to improve the accuracy of the then-most promising method of detecting an exoplanet (a planet outside the Solar System), known as radial velocity, Queloz pointed the 1.93-metre reflecting telescope towards the constellation of Pegasus.

This was just one of the many stars with which Queloz was calibrating the new system, but when the telescope fell upon 51 Pegasi it did something that every planet hunter had been dreaming about – it wobbled in a predictable and repeating manner. This old star, twinkling in the distant reaches of the night sky, was telling us something profound: it wasn't alone.

The method Queloz and Mayor were using to try to hunt down an elusive exoplanet at first sounds almost implausible. Radial velocity has nothing to do with measuring the presence of an exoplanet directly, as this is impossible due to the overwhelming glare of a star that's so bright it leaves any planets lost in the shadows. But the discovery of Poltergeist had shown that we could detect an exoplanet indirectly by looking for specific disturbances in the star's behaviour. A principle that the radial velocity method utilises with the fine-tuned observation of a target star.

RADIAL VELOCITY METHOD
The radial velocity method is based on the detection of variations (or 'wobbles') in the velocity of a central star, determined by the changing direction of the gravitational pull from an exoplanet as it orbits the star. The star's spectrum of light is redshifted as it moves away from Earth (left), and when it moves towards Earth, then it is blueshifted (right).

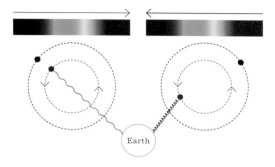

THREE ORIGIN THEORIES FOR HOT JUPITERS
Gas giants, or 'hot Jupiters', cannot form close to their parent star, and most likely form in one of three situations:
1. **Close in:** Planet forms near the star then travels closer in.
2. **Pulled in:** Planet forms far away from the star where gas giants are located, then as it interacts with gas and dust it pulls closer into orbit.
3. **Close encounter:** Planet forms far from the star, then is pulled away by an object such as a planet or comet, before stabilising close to the star.

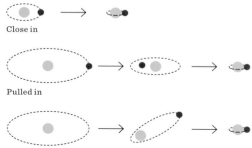

Close in

Pulled in

Close encounter

The method relies on a simple gravitational interaction that exists between every planet and star, including our own. In our solar system the vast mass of the Sun exerts its gravitational might on our planet's orbital motion as it does with all the planets and asteroids in the system. But gravity is never exerted in just one direction, and even though minuscule compared to the Sun, the relationship with the Earth also exerts a gravitational influence on the Sun's movement. This can be detected in the tiniest of wobbles in the motion of our sun caused by the gravitational pull exerted by the mass of our planet. Although negligible compared to the wobble caused by the more massive planets in our solar system, particularly Jupiter and Saturn, it is still measurable. This is called reflective motion and it's this tiny variation in a star's behaviour that Queloz and Mayor were trying to detect across all of those trillions of kilometres of space.

In principle they knew that a planet-induced wobble could be 'seen' by finding tiny measurable changes in the star's light spectrum or colour signature. Using a direct application of the Doppler Effect, the radial velocity method relies on the fact that a star moving away from the Earth will have its light 'stretched' and so its colour will shift and slightly redden from our perspective. When the same star 'wobbles' towards the Earth the light will be compressed and so shift slightly towards the blue end of the spectrum. These changes in a star's light are minuscule, but by meticulously measuring the amount of shift in the 51 Pegasi spectrum, Queloz and Mayor were able to measure it wobbling by just 53 metres every second – a wobble that told us this star had a companion, a planet that we could not only detect but even determine the size of. It was a true landmark moment in the history of science, and what would become a Nobel-prize-winning discovery that would change our view of the Universe forever.

51 Pegasi b, as it came to be known, was the first planet outside our solar system to be discovered around a sun-like star. But it was far from the alien world of our imaginations; this was no second Earth, no water world, no nurturer of life.

The wobble technique employed by Queloz and Mayor favoured the discovery of massive planets, capable of significantly distorting the star's light, so it was no surprise to discover that 51 Pegasi b was enormous. At around 150 times the size of the Earth, this planet resembles something far more like Jupiter than any of the terrestrial planets in our solar system. But it was this gas giant's proximity to its star that was truly surprising. In our solar system the gas giants orbit far from the Sun, with the terrestrial planets packed in much closer. It was the only planetary system we had ever been able to observe, so the assumption was this was a rule of thumb, an organisational principle for all planetary systems.

But 51 Pegasi b threw all of that out of the window. Here was a gas-giant planet approximately half the size of Jupiter orbiting closer to its star than Mercury orbits the Sun. Heating the atmosphere of this distant gas giant to at least 1,000 degrees Celsius, an atmosphere filled with blisteringly hot clusters of heavy elements and clouds not of water vapour but of silicates and iron. If we could get close enough, we would see a planet with an atmosphere so hot it would glow red, a planet half the mass of Jupiter inflated into a world even greater in volume than the giant of our solar system.

Travelling planets

When Didier Queloz and Michel Mayor first announced their discovery of 51 Pegasi b, many astronomers didn't believe it because the planet was in the wrong place. Here was a Jupiter-like world. It was enormous. It was massive, and yet it was parked right next to its star. At the time, astronomers and planetary scientists thought they understood planet formation and they fully expected that Jupiter, like other planets, would be far from their star because you needed to form those planets where it was cold enough that a lot of ice could build up the initial core, onto which the gas would pile and create this massive planet.

No one appreciated that planets could migrate after they formed, so the discovery of 51 Pegasi b really kicked off a deep understanding that, actually, after they form, planets in their respective planetary systems can move around and relocate.
David Charbonneau, Astrophysicist, Harvard Smithsonian Center

The discovery of 51 Pegasi b confounded everything we thought we knew about planets. We had explored our own star system and found that all the gas giants live far from the Sun in the 'outer' Solar System. We thought we understood how they evolved this way, built from the frozen icy rocks that can survive out past the snow line (far enough away from the Sun that volatile compounds such as water, ammonia, methane, carbon dioxide and carbon monoxide freeze into solids). Yet here was a Jupiter that raced around in an orbit once every four days – so close that it almost grazes the surface of its star.

It seemed 51 Pegasi b might be a fluke of nature, an anomalous discovery that would be sidelined as our search for more exoplanets gathered momentum. But as the discoveries began to trickle in through the 1990s it became clear that 51 Pegasi b was not entirely rare, and in fact this class of planet seemed to be something of a commonplace occurrence.

The discovery of a raft of hot Jupiter exoplanets revealed a new truth about the Universe. No longer reliant on a research sample of one, our solar system, we could now see that planetary systems did not simply conform to a structure of rocky planets close into the star and gas giants further out. It seemed from these first handful of discoveries that gas giants often end up migrating far from where they first formed. Dragged in towards their parent star by powerful gravitational forces, for some this ends up in a tortuously close orbit, for others it's a journey that ends in annihilation. We now know that even our own Jupiter began such a perilous journey towards the Sun, only to be pulled back from the brink by the counterbalancing force of its sister gas giant, Saturn. For all of its seeming stability, the story of our solar system and therefore the story of the Earth could have played out so very differently.

By the turn of the century it seemed as if the Universe might be awash with hot Jupiters, but as is often the case with science, jumping to conclusions can be a dangerous game. We now know that these types of planets are plentiful in our galaxy and thus are easier to find, but they are far from commonplace, perhaps making up around 1 per cent of all planets. The rush of discoveries in those early years was not down to the abundance of these planets but was instead a reflection of the biases of the techniques we were using to discover new worlds. Radial velocity and the other detection methods employed at that time favoured the discovery of large planets in tight orbits around their stars, and so by design it was hot Jupiters that at first appeared most frequently.

Slowly, with the development of new techniques, these giant planets gave way to a more eclectic mix of worlds, which is thanks largely to one extraordinary telescope, the greatest planet hunter of all time.

EXOPLANET TYPES

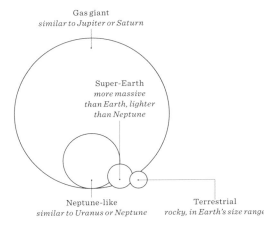

Gas giant
similar to Jupiter or Saturn

Super-Earth
more massive than Earth, lighter than Neptune

Neptune-like
similar to Uranus or Neptune

Terrestrial
rocky, in Earth's size range

THE
HUNTER

The launch of the Kepler Space Telescope on 7 March 2009 marked a profound moment in human history. Up until this point planet hunting was a side game for the world's most powerful telescopes, like the Spitzer and Hubble Space Telescopes, casting their eye to discover distant worlds in between a host of other mission objectives. But Kepler was different, the first space telescope designed and built entirely to hunt for exoplanets, and in particular Earth-like terrestrial worlds.

Launched by a Delta II rocket from Cape Canaveral Air Force Station, this 1-tonne telescope was propelled into an unusual 'Earth-trailing orbit'. This placed Kepler in an orbit around the Sun rather than the Earth to avoid the disruption that an Earthbound orbit can bring to a telescope's view of the Universe. With an orbital period or 'year' of 372.5 days, from the moment Kepler entered into its heliocentric orbit it began its slow fall away from the Earth, dropping farther behind our planet by approximately 26 million kilometres each year. On this trajectory, long after its retirement in 2018, Kepler will reach its farthest distance from the Earth in around 2035, when it will be on the other side of the Sun. It will then return to our local neighbourhood about 25 years later.

Above: Illustration of NASA's Kepler Space Telescope as it scans the skies above, searching for hidden exoplanets.

UNCOVERING HIDDEN WORLDS
The Milky Way galaxy between the constellations of Cygnus and
Lyra, targeted by the Kepler Space Telescope. The boxes show
the field of view of the telescope, containing over 100,000 stars.

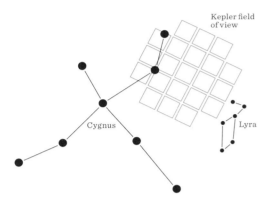

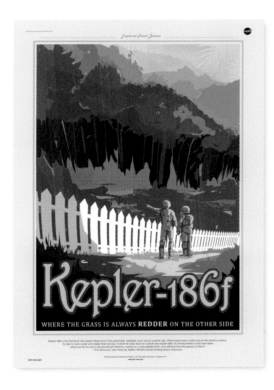

Above: NASA and JPL travel
poster advertising Kepler-186f, the
first Earth-size planet discovered
in a potentially habitable zone.

Although constructed around a 1.4-metre primary mirror, Kepler was not
designed for acuity but for sensitivity. Its aim was to deliver the maximum amount
of light to the single scientific instrument that it carries on board – an exquisitely
sensitive photometer that can continually monitor the precise brightness of up to
150,000 main-sequence stars. Intentionally defocusing the light from the primary
mirror, Kepler was not about capturing sharp images but maximising the
measurement of the intensity of light from its target stars. This in turn allowed
it to utilise a planet-hunting technique that enabled us to find Earth-sized planets
without actually seeing them.

The radial velocity method used in the detection of hot Jupiters like 51 Pegasi b
relies on the measurement of a change in wavelength of light, a wobble caused by
the gravitational disruption of a planet on its target star. But Kepler was designed
to use a different technique that relies on being able to use the subtlest of changes
in its brightness.

When you first hear about this principle, known as Transit Photometry, it's
almost impossible to imagine that such a technique can provide so much information
about tiny planets, specks of dust orbiting distant stars, but with the help of Kepler
it has been our most powerful weapon in the hunt for alien worlds.

The method relies on a simple principle, the fact that when a planet crosses or
transits in front of its parent star, the light from that star dips by a minuscule but
measurable amount. Using our solar system as an example of this phenomenon,
if you were measuring the light from the Sun and an Earth-sized planet transited
between you and the star, the light would dim by just 0.008 per cent. A Jupiter-sized
planet would cause a slightly larger dip.

Such changes seem almost inconsequential, but it is this dimming that Kepler's
photometer could use to measure the presence of tiny worlds hidden in the glare
of their stars. The method is far from perfect; such minute dips in light do result in
a high rate of false detection, which often require further verification, but that does
not detract from it being an incredibly powerful method for scanning the heavens
for other worlds. Not only can the transit method be used to detect the presence of a
planet, the level of dimming can also be used to estimate the planet's diameter, and
the frequency of this dimming can be used to calculate the duration of the planet's
orbit, or its year. Combined with an understanding of the characteristics of the
parent star, this can allow the temperature of the planet to be calculated, and so an
amazing amount of information can be gleaned from just the smallest twinkle of a
distant star.

Within a few weeks of being in orbit Kepler fixed its gaze towards the northern
constellations of Cygnus, Lyra and Draco, pointing towards a patch of sky chosen
so that the gaze of the telescope would never be interrupted by the blinding light
of the Sun. This target zone in the Milky Way was also chosen so that the stars
that Kepler would be observing would be a similar distance away from the galactic
core as our own solar system is, and also similarly close to the galactic plane. If the
Sun's position in our galaxy has influenced the chances of habitability, of life taking
a foothold on a terrestrial world, then looking into an area with a similar galactic
dynamic to our own perhaps might just increase our chances.

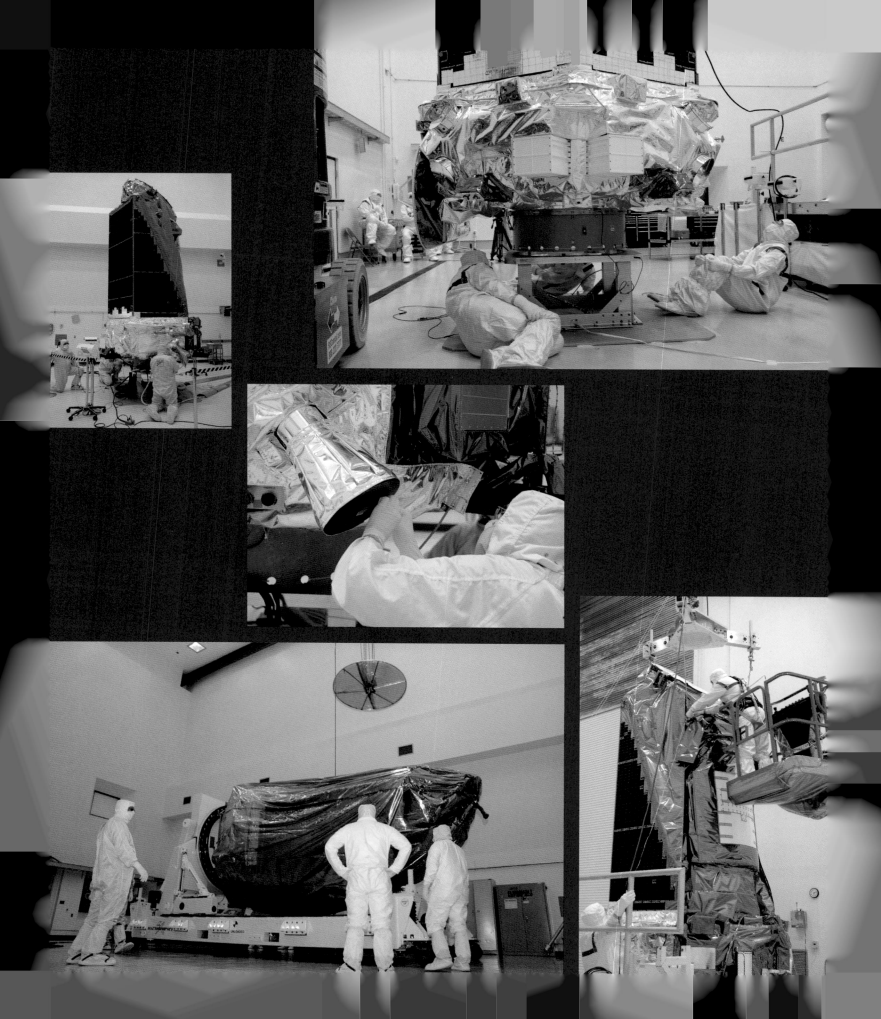

SEARCHING FOR SHADOWS

Phil Muirhead, Astrophysicist, Boston University

It is hard to understate how important the Kepler mission has been to our understanding of exoplanets. Prior to the Kepler mission, most exoplanet studies were discovering individual planets, one at a time. And we were really struggling to make broad inferences about the statistical nature of planets in the Universe. We needed more planets to make definitive statements about how common they are, how common Earth-like planets are, and how common Earth-like planets are in the habitable zones of their host stars. Kepler came along and blew everything out of the water, discovering thousands and thousands of extrasolar planets.

Kepler's method of discovering planets was a technique called the transit method, where you search for a shadow cast into space by the planet orbiting the star. Now to find planets via this method, you have to be lucky. The planets' orbit around the star must align such that the shadow is cast towards us. And that, unfortunately, has limitations, so just because a given star has a planet, it doesn't mean that Kepler would find it. If you look at enough stars, though, you're almost guaranteed to find a few with this chance alignment, and when you do, then all we can tell you about this planet is the orbital period. That is how long it takes to go around its star and the size of the planet. It does not tell you anything about the atmosphere of the planet, it does not tell you anything about the interior of the planet, and it also doesn't tell you necessarily the mass of the planet. For that, we would need to do follow-up observations; Kepler makes the discovery, it tells you the orbital period, and it tells you the size. And then astronomers take other telescopes, perhaps the Hubble Space Telescope, perhaps telescopes that are located on the ground on Earth – point at that system, and start to study it in other ways to try to determine the mass of the planet, the atmosphere of the planet.

So how special is Earth? We know now, thanks to research over the past few decades, that planets are actually quite common. One of the major results from NASA's Kepler mission is that planets exist around stars. So the fact that we have things like Mercury, Venus and Earth, Jupiter and Saturn, Uranus and Neptune, do not appear to be unique. We see planets like our planets orbiting other stars.

So Earth-like planets that orbit their stars at the same distance at which we orbit the Sun are very difficult to detect. They were right on the limit of what Kepler could find. And it took Kepler a long time to actually find those planets – you have to imagine they orbit their stars once every year. Kepler had to look for these and see that little dip once every year as they passed in front of their stars. Which made them very, very difficult to find. But even if you find a few you can start to do some statistics. You may not know the statistics very well, but at least you can start to ask those questions. And so even though we don't have very many Earth-like planets orbiting Sun-like stars, even though we don't know of very many, we're still able to calculate how common we think they are. And that is really wonderful news for questions about whether life exists on other planets.

THE KEPLER AND K2 MISSIONS

The approximate search areas for NASA's K2 and Kepler missions, both carried out by the Kepler Space Telescope.

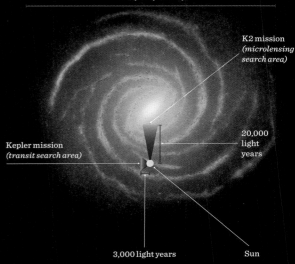

Milky Way Galaxy

K2 mission
(microlensing search area)

20,000 light years

Kepler mission
(transit search area)

3,000 light years

Sun

THE KEPLER TELESCOPE

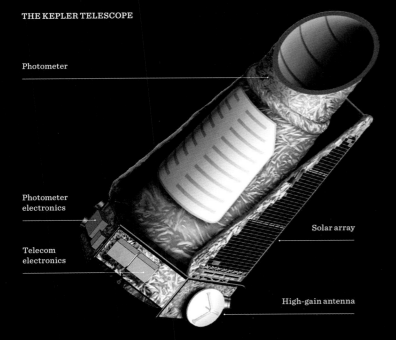

Photometer

Photometer electronics

Telecom electronics

Solar array

High-gain antenna

Opposite: NASA's Kepler spacecraft being prepared for testing at the Astrotech payload processing facility in Florida.

Alien life – science fact or fiction?
Any planet that Kepler has found in that so-called
habitable zone could have lakes or oceans on its
surface and could therefore host life. You might
think I'm talking about science-fiction, but there are
very, very reasonable conservative estimates that
are admittedly very, very error prone and uncertain
that suggest there are at least 10,000 advanced alien
civilisations in our galaxy alone. The caveat to all of
that is that we as a species will almost certainly never
know about any of it. If we get really, really, really
lucky one day, that would be the greatest discovery in
all of human history. But there are probably a pretty
jaw-dropping number of advanced alien civilisations
in our galaxy just because there are so many stars in
our galaxy alone.
*Grant Tremblay, Astrophysicist, Harvard
Smithsonian Center*

By the middle of May 2009, eight weeks after launch, all the tests and calibrations were complete and the telescope was ready to open its eye to the Universe, exposing its incredibly sensitive light meter to the light of 150,000 stars that had been selected for observation.

Designed to measure the light from all of these stars simultaneously, Kepler began the task of measuring their brightness every 30 minutes, searching for that minuscule dimming that would reveal a planet hiding in the glare. With each potential transit needing to be observed at least three times before it could be confirmed, the scientists waiting back home knew that at least to start with it would be most likely Kepler would add to the collection of hot Jupiters that we know of in the galaxy. Simply due to their size, these giant planets give a dip in light that is easier to check, and because they orbit tight in to their parent star, the time it takes for the planet to complete its orbit and transit in front of Kepler's field of view again is relatively short, and so faster to verify than a planet further out. And this is exactly what happened as the first data from Kepler began to trickle in throughout 2009. But everything about Kepler and its mission from its line of sight to the peak sensitivity of its photometer was designed to look for much smaller planets, Earth-sized worlds orbiting in the habitable zone of one of those distant stars. And so if you are trying to find a planet like Earth and you need it to orbit three times before you can be sure of its detection, it was expected to take three years or longer for Kepler to start returning its most precious results.

Slowly but surely, Kepler began returning data to the team eagerly waiting on Earth. First it was just a trickle of candidate planets, then a few hundred and then a cascade, and by the end of 2011 Kepler had identified well over 2,000 planetary candidates, ranging from Earth-size worlds to planets larger than Jupiter.

Once a candidate planet had been identified, a fleet of ground-based telescopes could then be deployed to train their gaze on the relevant star and attempt to confirm the finding by repeating the transit measurement or by using another method of detection, like radial velocity. One of these candidate planets was confirmed a few months later and named Kepler-36b. Orbiting around a subgiant star (a star that is redder and larger than main-sequence stars of the same luminosity) 1,200 light years from Earth, this was a planet that at first glance appeared enticingly familiar.

Not too big and not too small, this was one of the first rocky planets to be confirmed to exist outside our solar system. In orbit around a living star similar to our own, here finally was a planet that at first glance appeared recognisable, a planet that was in some ways like our own home. Analysis of the photometry has revealed Kepler-36b to be substantially bigger than the Earth, measuring 1.5 times the radius and 4.5 times the mass of our home planet – a set of characteristics that defined it as a super-Earth. But Kepler-36b was no second Earth. As the data began to reveal more detail about its orbit, its size and its temperature, it became clear that this rocky planet was anything but habitable, a tortured soul in eternal torment around its star.

KEPLER-413B BINARY SYSTEM
This diagram shows Kepler-413b's 66-day orbit, which is unusually tilted 2.5 degrees to the plane of the orange and red dwarf binary stars.

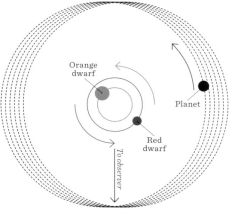

Orange
dwarf

Planet

Red
dwarf

To observer

56,327,040 kilometres

Opposite: NASA's Kepler
mission has discovered a
cold, uninhabitable planet,
called Kepler-16b, with two
suns instead of just one.

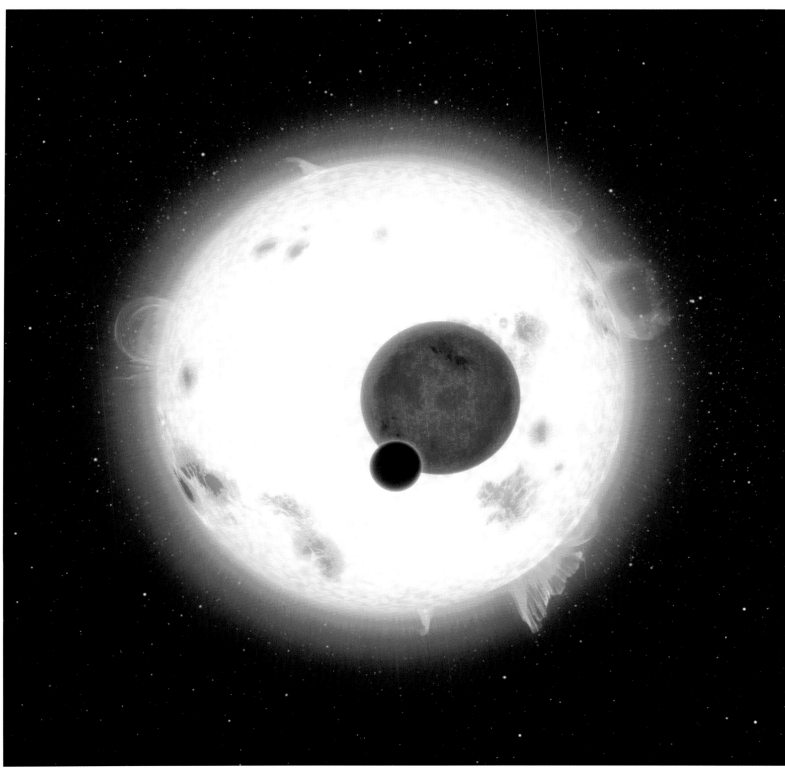

Earth   Kepler-62f   Kepler-62e   Kepler-69c   Kepler-22b

KEPLER-36

Phil Muirhead, Astrophysicist, Boston University

Kepler-36 is a very exciting system because you have two planets that get very, very close to each other as they orbit the star. Gravity affects all mass, all matter, all things with mass; all things that are made of matter are pulling on each other and when you get two planets really close together, they start pulling on each other, too. So it's hard to imagine how you get the system of two planets orbiting a star and orbiting such that they're so close and how the planets evolve over time such that they end up in this particular scenario.

Kepler-36b and Kepler-36c are also very different planets. One is terrestrial and rocky; the other is much more like a gas giant planet. It's very low density. So they likely didn't form near each other, but somehow ended up near each other over time. And that's very, very interesting. Also, they're both tidally locked to their host star. This means that they are orbiting so close that the gravitational force of the star is higher on the facing side of the planet, which keeps the same side of the planet always facing the star. On Earth we experience tides primarily from the Moon, and those tidal forces cause the oceans on the Earth to swell and change. But our impact

on the Moon is much stronger, forcing the Moon to rotate around its axis in the same amount of time as it rotates around the Earth, which is why we always see the same side of the Moon. This is called synchronous rotation. In solar systems with stars much larger than our sun, or with planets sitting much closer to the central star, the star can hold planets in synchronous rotation. Kepler-36b and -36c, for example, are so close to their host star that the tidal force from the star forces the planets to rotate with the same period as its orbit.

A consequence of this tidal locking is that one side of these planets experiences rays from the star constantly while the other side of the planet is looking at the night sky all of the time. So you have this massive difference in temperature – the side of the planet facing towards the star gets very, very hot while the side of the planet facing away from the star gets very, very cold. If you have some atmospheric circulation, ie winds, that will help ameliorate the problem, but still, you'd have this massive temperature difference. And that's a very exotic situation compared to what we experience on Earth.

TIDAL LOCKING

A planet is known as tidally locked when it is so strongly affected by the gravity of its star that it must rotate synchronously, thus it always has the same face pointing towards the star.

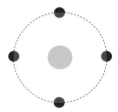

Kepler-36b orbits just 17 million kilometres out from its mother star, far closer than the Earth is to the Sun (average 150 million kilometres) and far closer than even the innermost planet in our solar system, Mercury (at an average distance of 58 million kilometres). Speeding around on its orbital path it completes its 'year' every 13.8 days. And it orbits so close to its parent star that it has become locked into a rotational trance, gravitationally bound so that one side of the planet faces permanently inwards while the other side faces permanently out – a dynamic that has created a world of mind-boggling extremes.

On the star-facing side of the planet, the glare of the star creates a punishing heat with surface temperatures in excess of 700 degrees Celsius. While Kepler-36b's opposing face is permanently turned away from the fierce star, its face pointing into the darkness of space, its surface temperature never rises above minus 100 degrees Celsius, a freezing-cold world shrouded in eternal darkness.

Imagine a world where the Sun stays in the same place in the sky forever. One side of the planet is in eternal night; the other side in eternal daylight. And even that twilight strip between day and night we think will suffer from extreme conditions.

Kepler-36b demonstrates that there is so much more to being a habitable world than the size and nature of the planet itself. There are the details of its star and the details of its orbit around the star that massively impact on the characteristics of any world, but that's not where the factors influencing a planet's personality end.

We now believe that many star systems – perhaps even the majority – are multi-planetary, and the Kepler-36 star system is no exception. The rocky super-Earth 36b is not alone around this star – it has a much larger sibling orbiting within remarkably close proximity. Kepler-36c is a gas giant of a planet, eight times the mass of the Earth and almost four times the radius, it orbits just 19 million kilometres away from the star, taking 16.2 days to complete its year. The two planets (that we know of) in this system are in such close proximity that every 97 days they come within 2 million kilometres of each other, a distance between these massive worlds that is just five times the Earth–Moon separation.

It's this close proximity that violently transforms the already battered landscape of 36b. The inner planet 36b completes seven orbits in the time it takes the outer planet to complete six orbits, and so with a 7:6 orbital resonance it means that every 97 days the gas giant looms large in the otherwise static skies of 36b. The gravitational impact of this encounter on the terrestrial world is dramatic, pulling and stretching the core of the planet, driving great tidal forces through the very rock it is made of until it erupts in intense volcanism. The result is a molten world, where the air is choked by dust and gas and the ground pelted by lava and rock. Where the surface is crisscrossed with rivers of lava freezing on the cold dark side and endlessly smouldering on the bright side of the planet that is too harsh for life to survive and too hostile to be a home. Worlds like Kepler-36b highlight just how complex a planet's story can be.

At the time of writing, in summer 2021, we have discovered 4,352 planets outside of our solar system in the Milky Way. Over 1,300 of these are gas giants, massive planets like 51 Pegasi b, composed mainly of helium and hydrogen that often orbit close-in to their star. Even more of them, nearly 1,500, are classified as Neptune-like planets, similar in size to their solar-system namesake, with vast hydrogen- and helium-dominated atmospheres but with heavy metal cores. Then there are the

Opposite: An artist's impression of how Kepler-36c might appear as viewed from the surface of neighbouring Kepler-36b.

Above: Our moon is tidally locked to the Earth, so no matter where we are on our planet's surface, we will only ever see one side of it.

'The Universe is not only queerer than we suppose, but queerer than we can suppose.'
J.B.S Haldane, Biologist

super-Earths like Kepler-36b, a class of planet unlike any other in our solar system, worlds that are considerably more massive than the Earth (up to ten times the mass) but much smaller than the ice giants like Neptune and Saturn.

Over 1,300 super-Earths have been discovered so far, suggesting they are a common feature of other star systems. The name itself can be misleading, though, as these are not necessarily Earth-like worlds; they are a mysterious category of planet with characteristics unlike anything we have ever directly explored in our own solar system. That leaves just over 160 planets that we class as terrestrial, worlds that, like Mercury, Venus, Earth and Mars, are rocky worlds – not too big and not too small. Kepler was responsible for discovering the vast majority of these worlds: 140 of them were found by the space telescope.

But for all of the hope that Kepler's exploration of the galaxy brought with its discovery of these terrestrial worlds across the Milky Way, it has also brought an overwhelming realisation that each of those terrestrial planets, in fact each of the 4,000-plus planets that we've discovered to date, is different from all the others, an alien and exotic group with no planet identical to any of the planets in our solar system. There are too many different factors that add up to make a planetary system for there to be mirror images of our own solar system scattered right across the Milky Way.

It seems our first steps towards exploring the alien worlds beyond our own have revealed a deeper truth about the Universe.

Although the laws of nature that form planets are the same everywhere and the fundamental ingredients of these planets are simple and the same, the nature of a planet also depends on the history of its formation, and the environment around its parent star out of which it formed. Every planet has a different story to tell and this wholly unexpected but exciting discovery certainly complicates the search for life.

There will be no exact replica of Earth anywhere in the Milky Way, so as we continue to hunt for alien life we must redefine what we are looking for in order to understand what, at its most fundamental, turns a planet into a home.

Of the 4,352 alien worlds we've discovered so far, Kepler, our most tenacious planet hunter, discovered over 60 per cent of them – a total of nearly 2,500 planets. It's a staggering number of discoveries for a single telescope, but perhaps even more extraordinary is what this discovery tells us about the abundance of planets in the rest of the galaxy. All of the 2,414 planets that Kepler found are located in just one patch of the sky, a relatively tiny scrap of the Milky Way that covers around 0.25 per cent of the sky. To explore the whole of the night sky in the same depth would take 400 Kepler telescopes, but as wonderful as that sounds, in a sense we don't need them. By using the Kepler data as a guide to the frequency of planets in the Milky Way we can extrapolate out from the one patch we have explored in minute detail to start seeing the bigger picture. If we take the patterns it found – most importantly that every star in the galaxy is joined on average by at least one planet – and multiply that out to the rest of the galaxy, we can confidently predict that there are at least 100 billion planets in the Milky Way and probably many more – perhaps there are even more planets than stars.

Kepler revealed to us an overwhelming truth: our galaxy is crowded with an inconceivable number of worlds, planets with an endless variety of characteristics and chemistry. So, if we are to look for life amongst this infinite array of worlds, what are the essential characteristics we are looking for, what are the minimum requirements to make a planet a viable world that can generate and support the existence of a living entity? The only answers we have to these fundamental questions come not from looking outwards to the galaxy but from looking inwards at the only one of the billions of planets in the Milky Way that we have been able to investigate. Our understanding of life comes from a sample of one, from looking at our living planet and understanding amid all of the complexity, what is the essence of life and therefore the likelihood of it occurring elsewhere in the Universe. Only with this knowledge can we begin to narrow the search for the living worlds that the simple numbers tell us must be somewhere out there.

Left: Wistman's Wood, Dartmoor, is an ancient woodland that exemplifies the Earth's ability to sustain life through generations.

Above: Sea anemones clinging to the stones in a rock pool illustrate the adaptability of life on Earth in order to survive.

ALIEN WORLDS

In February 2021 a group of scientists from the British Antarctic Survey made an extraordinary discovery. Drilling down through almost a kilometre of the Antarctic ice shelf, the team were attempting to take a rock sample, a sedimentary core from the floor of the south-eastern Weddell Sea that lay beneath this vast amount of ice. But as the team broke through the bottom of the ice sheet and into the freezing waters below, their attempt to drill into the sea floor came to an abrupt halt. Something was in the way.

The Filchner–Ronne Ice Shelf, like all ice shelves, was formed when a glacier flowed down a coastline, with the momentum of the glacier creating a large body of ice floating on top of the ocean's surface that became hidden hundreds of metres below. Because of the way they are formed, ice sheets often carry large boulders that have been picked up by the glacier as it moves across land before it deposits them on the sea bed. In what at first seemed like bad luck, this type of boulder is exactly what the British scientists' drill hit as they approached the sea floor, thwarting their attempt to take a sample of the sea bed and seemingly wasting all their exploratory work.

However, before moving on the team sent a camera down the hole, into the black, icy water, to take a look at exactly what was blocking their route. As the images came back to the surface they were in for a shock. They had indeed hit one of those stray boulders, but the images showed this rock was far more interesting than they could have possibly imagined. Here, 500 metres below the base of the ice shelf, 260 kilometres away from the nearest open water, sitting in total darkness and in temperatures of -2.2 degrees Celsius, there was an abundance of life thriving on this barren rock. The images revealed at least two types of sponge, tube worms and stalked barnacles all clinging to the rock in an environment that had been considered incompatible with this kind of life.

As one of the science team, Dr Huw Griffiths, said after the find: 'It's slightly bonkers! Never in a million years would we have thought about looking for this kind of life, because we didn't think it would be there.'

Here is a complex community of living organisms stuck on a rock deep below the base of the ice shelf, a vast distance from any direct food source, and yet it is thriving. We don't know what they feed on, perhaps dead plankton washed here by the powerful currents underneath the ice, nor do we know how long life has existed here or if they are new species. 'This is by far the furthest under an ice shelf that we've seen any of these filter-feeding animals,' said Griffiths. 'These things are stuck on a rock and only get fed if something comes floating along.'

The discovery is just one of the latest examples of the endless adaptability of life on Earth. Again and again over recent years our expectations of the limits of habitability have been overturned, with life found not just existing but thriving in all manner of extreme environments.

From deep beneath the Antarctic ice shelf to the upper reaches of the atmosphere, spanning arid deserts and toxically acidic volcanic lakes, our exploration of Earth has taught us that life is remarkably adaptable, tenacious and resilient. Earth is home to at least 9 million living species – all animals, plants, fungi, protozoa and bacteria are related to each other in the great web of life that stretches back to a single ancestor that emerged on our planet almost 4 billion years ago. These endless forms occupy just about every available niche on the planet, including places we once thought too harsh and hostile for life.

Above: The astounding discovery of life beneath the Filchner-Ronne Ice Shelf.

Opposite: The harsh waters of Lake Natron, a salt lake in Tanzania, surprisingly teem with life, including salt-loving microorganisms and flamingos.

'Our top best chance for finding life in our solar system beyond Earth, is probably the planet Mars. There's almost certainly not life there now, but there was certainly liquid water on Mars at one point. And there maybe was a very simple form of life there some time in the past.'
Grant Tremblay, Astrophysicist, Harvard Smithsonian Center

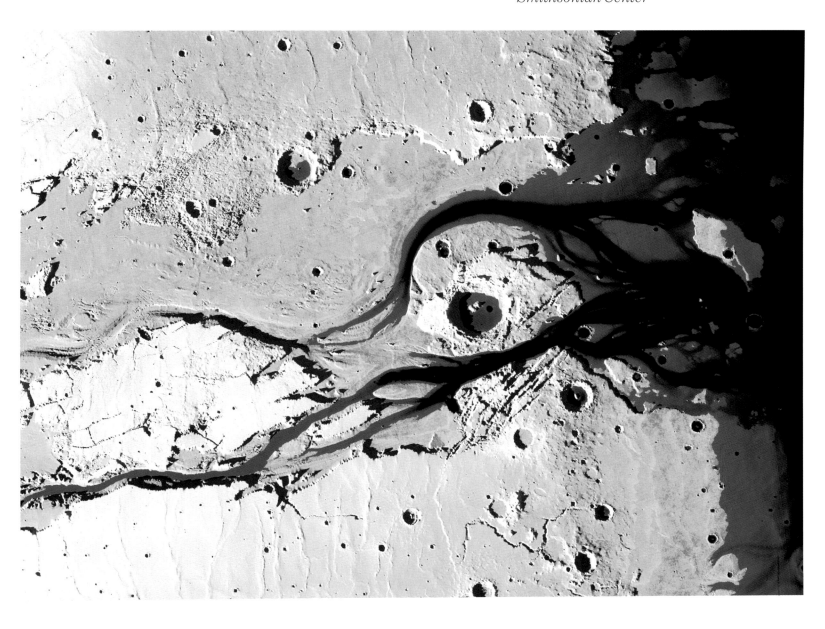

Opposite: River systems like the Betsiboka River in Madagascar are part of an essential, life-giving ecosystem on Earth.

Above: The mouth of the dried-up river that formed the Kasei Valley on Mars 3.5 billion years ago, before it became a desert world.

THE BUILDING BLOCKS OF LIFE
The key elements found in all life on Earth – carbon, nitrogen, oxygen and iron – shown here as atomic diagrams.

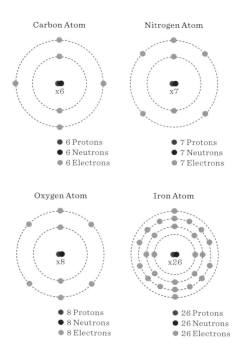

Carbon Atom

x6

- 6 Protons
- 6 Neutrons
- 6 Electrons

Nitrogen Atom

x7

- 7 Protons
- 7 Neutrons
- 7 Electrons

Oxygen Atom

x8

- 8 Protons
- 8 Neutrons
- 8 Electrons

Iron Atom

x26

- 26 Protons
- 26 Neutrons
- 26 Electrons

Everything we know about life we've discovered by looking at the living world around us, and by studying life on Earth in so many different environments we've been able to understand not just the limits of life but the minimum requirements for survival and the conditions that we think were necessary for the first life on our planet to emerge.

Now you might legitimately ask the question, can we transfer the knowledge we have of life here on Earth to every other planet in the Universe? I would say emphatically yes, because the laws of nature are universal. The laws of physics and chemistry that underpin biology here on this planet will apply to every planet out there in the Universe, whether we've discovered them or not.

For all of the endless variety of life on our planet, we can actually boil down the chemistry of life to just a few universal ingredients. In terms of the elemental building blocks, life needs not much more than carbon, nitrogen, oxygen and iron. It also requires a ready supply of energy. Here on Earth that energy comes from just two different sources: geothermal energy (the heat within the core of our planet that's left over from the vast collisions that occurred during its formation and the radioactive decay that continues deep within our planet to this day), but first and foremost from starlight, from the vast amounts of energy that pour across

WATER ON EARTH

The vast majority of the Earth's water is contained in oceans. The freshwater lakes and rivers, which we drink and depend on for survival, make up only a fraction of the water on Earth.

Total global water

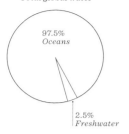

97.5%
Oceans

2.5%
Freshwater

Total freshwater

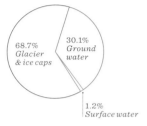

68.7%
Glacier & ice caps

30.1%
Ground water

1.2%
Surface water

Total surface water

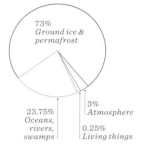

73%
Ground ice & permafrost

23.75%
Oceans, rivers, swamps

3%
Atmosphere

0.25%
Living things

What makes a water world?

In the case of Earth, we have an abundance of water on the surface, which is extremely important for life. In the case of the other planets, we don't see liquid water on their surfaces, on planets that are closer to the Sun, it is just too hot. If you were to take a bowl of water and put it on the surface of Venus, it would boil and turn to gas almost instantly. If you were to take a bowl of water and put it on the surface of Mercury, it's going to boil off. Mars is quite cold, and while there is no liquid water on the surface of Mars currently, there is evidence for it in the past, and we know this thanks to these fantastic NASA missions to Mars, studying its surface and the evolution of its atmosphere. There's evidence for water on Earth going back billions and billions of years. In fact, we have evidence for life on Earth for billions of years. So there's something about our planet that allowed water to stick around.

There are a lot of interesting proposals for what happened. Mars is much smaller than the Earth and so the gravity is much lower, has much less gravity. *Phil Muirhead, Astrophysicist, Boston University*

the Solar System and bathe this planet in sunlight. Beyond this there is one other fundamental ingredient that we are certain all life depends upon – liquid water.

Water is so ubiquitous in our lives that we tend to think of it as the simplest of substances, but liquid water and the role it plays in life is deceptively complicated. First and foremost, water is a very powerful solvent, capable of dissolving more substances than any other liquid, which is why it is often referred to as the 'universal solvent'. This capability to dissolve a wide variety of molecules makes it an invaluable component, facilitating structure, function and transportation within all living things. As well as being a powerful solvent it also has complexity in its own structures, constantly forming and disappearing, and they act as a kind of scaffolding around which biology can happen, orientating organic molecules so that they can react together. All of this function adds up to make liquid water not only important for life but seemingly a pre-requisite for its very existence. Every living thing that we know of here on Earth requires liquid water to survive.

So if we are to hunt down life out there in the galaxy, it's a good assumption that it will be life forms that are dependent on liquid water. And that is exactly what has informed our strategy. As NASA and other explorers set out to search for potential habitats for alien life, they all do so under the mantra 'follow the water'.

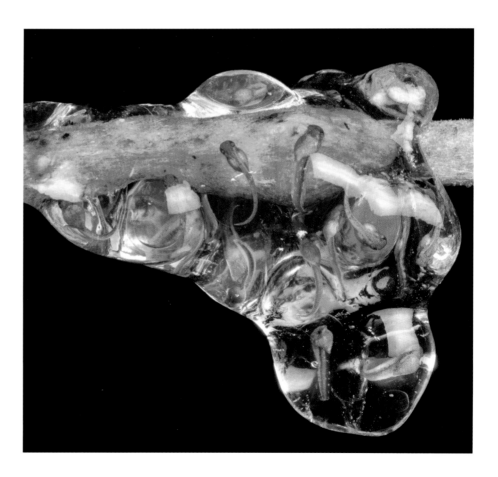

Opposite: The flow of water sculpts land and ice and is an essential life-giving element for myriad life forms.

Above: The eggs in this Crump's treefrog nest develop mostly in water, demonstrating water's life-sustaining properties.

PERSEVERANCE

Landing on Mars on 30 July 2020, Perseverance is the most sophisticated rover that NASA has thus far sent to the red planet. Its purpose is to find signs of ancient life on the planet, and to help it to find this evidence, the rover is equipped with the most cutting-edge technology.

SuperCam
An instrument that can provide imaging, chemical composition analysis and mineralogy. The instrument can detect the presence of organic compounds in rocks and regolith from a distance and also has a significant contribution from the Centre National d'Etudes Spatiales, Institut de Recherche en Astrophysique et Planétologie (CNES/IRAP), France.

Mars Environmental Dynamics Analyser (MEDA)
A set of sensors to provide measurements of temperature, wind speed and direction, pressure, relative humidity and dust size and shape.

Radar Imager for Mars' Subsurface Experiment (RIMFAX)
A ground-penetrating radar to provide centimetre-scale resolution of the geologic structure of the subsurface.

Mars Oxygen ISRU Experiment (MOXIE)
An exploration technology investigation to produce oxygen from Martian atmospheric carbon dioxide.

PERSEVERANCE'S SCIENCE OBJECTIVES

Geology:
Study the rocks and landscape at its landing site to reveal the region's history.

Astrobiology:
Determine whether an area of interest was suitable for life, and look for signs of ancient life itself.

Sample Caching:
Find and collect promising samples of Mars rock and soil that could be brought back to Earth in the future.

Prepare for Humans:
Test technologies that would help sustain human presence on Mars someday.

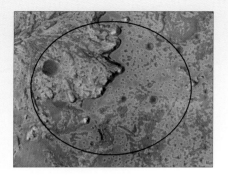

Mastcam-Z
An advanced camera system with panoramic and stereoscopic imaging capability and the ability to zoom. The instrument can also help scientists assess the mineralogy of the Martian surface and assist with rover operations.

Scanning Habitable Environments with Raman & Luminescence for Organics & Chemicals (SHERLOC)
A spectrometer to provide fine-scale imaging and use an ultraviolet (UV) laser to determine fine-scale mineralogy and detect organic compounds. SHERLOC is the first UV Raman spectrometer to fly to the surface of Mars and will provide complementary measurements with other instruments in the payload.

Planetary Instrument for X-ray Lithochemistry (PIXL)
An X-ray fluorescence spectrometer with high-resolution camera to determine the fine-scale elemental composition of Martian surface materials. PIXL will provide capabilities that permit more detailed detection and analysis of chemical elements than ever before.

FOLLOW
THE WATER

'Our solar system is far richer in the potential and opportunity to host life than we ever expected. There's a concept: follow the water; where there's water, there might be life. And we have found so many solar system bodies with water – Jupiter's icy moons, especially Europa, then there are Saturn's moons, including Enceladus. There is even one of Saturn's moons, Titan, which doesn't have water, but it has liquid ethane and methane lakes.'
Sara Seager, Professor of Planetary Science, Physics, Aeronautics and Astronautics, MIT

The Universe is awash with water and has been since the dawn of time. We've seen the signature of its formation in the dust clouds of dying stars, planetary nebulae like the Helix Nebula that have seeded the Universe with huge amounts of water since early in its life. We've been able to witness the formation of water in vast amounts in interstellar clouds like the Orion Nebula, where we have measured the activity of these vast cosmic factories and seen that in a single day Orion generates enough water molecules to fill the Earth's oceans 60 times over. Everywhere we look we see water molecules in vast amounts, but although it is one of the most abundant molecules in the Universe, all of this water is almost entirely locked away in frozen dust clouds that float aimlessly around. It's only when these grains of interstellar dust get wrapped up in the formation of a galaxy and a star system that things really start to get interesting.

Five billion years ago the cloud of gas and dust that would go on to form our solar system was full of water. Swirling in the frozen dust cloud around the newly formed star, it would find itself incorporated into all eight of the planets and dozens of moons that would form around our sun. Yet among our planetary neighbours Earth was the only world to hang on to its water in liquid form. Orbiting at a distance from the Sun – where it is neither too hot nor too cold – and holding on to its protective and stabilising atmosphere, Earth was able to form and maintain oceans, rivers and lakes. And it was here, just half a billion years after its birth, that life began. Triggered by the fertile combination of geothermal energy, in contact with high concentrations of chemicals and minerals all brought together by the magical solvent – liquid water.

This is the first and only time that we know of in the history of our galaxy and beyond, that the conditions conspired to transform chemistry from the inanimate to the animate. But we know this cannot be the only time amongst the billions of planets out there that these conditions have occurred. We know the galaxy is full of rocky worlds, worlds that must be rich in minerals and by the very nature of their formation have energy bursting to the surface from deep within, but how many of these are water worlds and how many of them can we find?

On 30 October 2018 NASA announced that the Kepler Space Telescope was dead. Awakened from its sleep mode one last time, Kepler had peered into the darkness and begun collecting scientific data for its nineteenth observation cycle. But now, after a mission longer and more successful than anyone could have imagined, this great explorer was out of fuel, unable to hold its position and destined to continue its orbit in perpetual silence. But this was far from the end of Kepler's contribution to our search for alien worlds. The data Kepler provided in its nine and a half years of operation continues to provide new insights and directions in our search for worlds with the potential to support life.

Opposite: Potassium and sodium deposits in Lake Assale, Ethiopia, demonstrate how the combination of liquid water, chemical elements, minerals and geothermal energy make life possible on Earth.

Above: The Helix Nebula actively produces water molecules and propels them out into the galaxy.

KEPLER'S WORLDS
This chart plots the data from the eighth Kepler planet catalogue – the most recent discoveries are in yellow. Size and orbital period (both relative to Earth) are on the Y and X axis respectively. Planets in the bottom half are likely to be rocky, while the top half are likely gas giants. Along the middle, there are ocean worlds and ice giants.

● Previously known   ● New

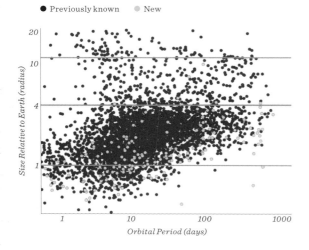

The chart on this page shows the data from the eighth Kepler planet catalogue released in August 2017. It contains the 4,034 planetary candidates discovered by Kepler. We know that 2,414 of these candidates have been confirmed as planets but there is still much to explore in terms of this data and the worlds it is pointing us towards. What's particularly striking about this chart is that it places all of Kepler's candidate planets within a Goldilocks framework, revealing the likelihood of liquid water existing on any of these hundreds of worlds.

Along the vertical axis each candidate planet is plotted according to its size relative to Earth – the property that determines the kind of atmosphere it might have. Too large and it will almost certainly be some kind of gas giant, too small and it perhaps won't be able to hold onto the atmosphere that life needs to drive a water cycle and protect that water from the hostile environment of space. On the horizontal axis the candidate planets are plotted according to orbital period, an indication of how far the potential planet is from its star, and so an approximation of how much heat it is directly receiving. This tells us where it might sit in the magical Goldilocks zone – whether it is either too hot or too cold for water to exist as a liquid. What's clear from this plot is that the majority of Kepler's finds are weird and wonderful worlds not anything like the Earth. The vast majority are much bigger and closer to their star and so unlikely to be able to hang on to liquid water and its life-giving properties.

But we know that this data cannot be taken as an accurate reflection of the frequency of Earth-sized planets orbiting in a habitable zone. Even with Kepler's extraordinary sensitivity it was much easier for it to find large planets, especially those with short orbital periods and so close to their star. This substantial observation bias means that the frequency of Earth-like planets is almost certainly lower in this data than in reality. And so it's likely there are far more Earth-like planets out there just waiting to be found, requiring the next generation of planet-hunting technology to track them down.

Even with this disadvantage Kepler has still provided a clutch of potential planets that look – at least at first glance – potentially Earth-like. Around 30 worlds that have the size, atmosphere and temperature that might possibly allow liquid water to accumulate on their surface. Planets that look teasingly plausible as habitable worlds, worlds that we can target and on which we can begin to look for evidence of both water and life, but to delve deep into these distant worlds would require more than just the Kepler data. Even a planet hunter as successful as Kepler needs some assistance.

Above: Dust and gas emissions from comet 67P/Churyumov-Gerasimenko, photographed by the probe Rosetta, which would go on to land on the comet in 2015.

Opposite: Artist's concept of Kepler-186f, the first validated Earth-size planet to orbit a distant star in the habitable zone.

KEPLER'S TIMELINE

6 March 2009: NASA's Kepler Space Telescope launched on a three-stage Delta II rocket, from Cape Canaveral Air Force Station in Florida.

8 April 2009: Kepler awakens in a star-studded patch of sky in the constellations Cygnus and Lyra. Out of the 4.5 million stars, the telescope would almost continuously monitor more than 170,000 for dips in brightness as planets cross in front of their parent stars.

4 January 2010: Kepler's first five planetary discoveries, named Kepler-4b, -5b, -6b, -7b and -8b, are hot Jupiters – gas-giant infernos – orbiting their parent stars in just days, with surface temperatures above 1,000 degrees Celsius.

10 January 2011: The first solid evidence of a rocky planet comes into view with Kepler-10b, a 'lava world' orbiting so close to its star

that its star-facing side could be an ocean of molten rock. Kepler discovers hundreds of rocky planets.

15 September 2011: While Kepler-16b experiences double sunsets, it turns out to be a gas giant that might lack a solid surface.

5 December 2011: Kepler-22b is the first planet found in the 'habitable zone'. At more than twice Earth's diameter, it is among the most common planet sizes.

18 April 2013: Kepler's discovery of a new planetary system brings into view three 'super-Earth' size worlds in the habitable zones of their parent stars – Kepler-62e and -62f, and Kepler-69c (though subsequent studies have shown that Kepler-69c no longer fits this category). The discoveries prove that small planets in the habitable zone of their parent stars exist beyond our solar system.

14 May 2013: The Kepler spacecraft loses its second reaction wheel, bringing an end to science observations. The mission team searches for another way to operate the telescope in order to restart science observations.

17 April 2014: Kepler data reveals the first Earth-size planet in the habitable zone of its star. Kepler-186f orbits a cool, red dwarf star about 580 light years away. Its discovery is a step closer to finding Earth-like worlds.

May 2014: Kepler begins a new mission as 'K2' using a technique to take advantage of the pressure of sunlight to help stabilise the telescope's pointing. This requires the spacecraft to switch its field of view every three months, bringing many new patches of sky under Kepler's gaze.

23 July 2015: The Kepler team discovers a super-Earth 60 per cent larger than our

planet. Kepler-452b orbits in the habitable zone of a Sun-like star and is considered the closest found to an Earth-Sun analogue. A recent study disproves this.

21 October 2015: Kepler's extended K2 mission uncovers evidence of a small, rocky planet being torn apart as it spirals around a white dwarf star, allowing astronomers to witness the final stages of a solar system.

January 2016: Named after astronomer Tabitha Boyajian, Kepler picks up strange fluctuations in its brightness, found to be a dust cloud moving around 'Tabby's star'.

10 May 2016: The Kepler mission hauls in more than 1,200 exoplanets; many could have an Earth-like composition.

22 May 2017: Using data from Kepler's extended K2 mission, astronomers pin

down the orbital period of the outermost planet in the TRAPPIST-1 system – home to seven Earth-size planets. The data backs up the theory the planets migrated inward during the system's formation.

19 June 2017: The most comprehensive catalogue of exoplanets from Kepler's first four years of data includes 4,034 candidate planets, with 2,335 confirmed. The initial estimate of near Earth-size, habitable zone planets was 30. Since then, new data and preliminary analysis shows the actual number is possibly between 2 and 12.

14 December 2017: An eighth planet is found in the Kepler-90 system, equal to our own solar system in having the largest number of known planets. All crowd closer to their star than Earth to our sun. The discovery is made, in part, using artificial intelligence.

11 January 2018: An Australian car mechanic discovers a four-planet system with Neptune-size worlds within the data. Scientists find a fifth planet.

18 April 2018: Kepler's successor, the Transiting Exoplanet Survey Satellite, rockets into space. TESS uses the technique pioneered by Kepler – watching for dips in starlight – to find exoplanets orbiting bright, nearby stars.

May 2018: Kepler finishes six months of observing supernovae, capturing the beginning stages of these stellar explosions with unprecedented precision to resolve the mystery: What sets them off?

30 October 2018: Kepler completes its nine-year odyssey observing more than half a million stars. Kepler's discoveries have taken us one giant step further in our search for life in the galaxy.

THE SMELL OF RAIN

In 2015 Kepler pickled up the flicker of a planet crossing in front of a star 124 light years from Earth in the constellation of Leo. This was not by a long way the smallest of worlds that Kepler had found. It was about twice the radius and eight times the mass of our planet, putting it in the super-Earth category of exoplanet discoveries, a world where the increased gravity would dwarf the forces we see play out on our own planet. This world wasn't orbiting a sun-like star either; its parent star was a highly active and volatile red dwarf, the smallest and coolest type of main sequence star. These long-lived stars are the most numerous type of star in the galaxy, but they are not necessarily the most conducive to the creation of home. But for everything that counted against this world being a hopeful target in our search for alien life, there were also some intriguing other details in the data.

Reaching 21 million kilometres from its star and completing one orbit in just 33 days meant this world was close to its star and almost certainly tidally locked, with one face forever facing inwards. But around a red dwarf star such close proximity does not necessarily mean intense heat; in fact the data suggested that this planet, named K2-18b (because it was the eighteenth planet discovered during the 'second light-K2' part of the space telescope's mission) was potentially in the habitable zone of this cool star with a likely surface temperature hovering around zero degrees Celsius. And that was not all that made this world stand out from much of the Kepler data. A planet circling in the habitable zone of a red dwarf like this opened up the possibility of not just observing the planet, but with much less glare from the star it might be possible to explore the atmosphere of this world as well.

LEO

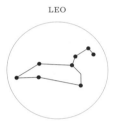

INFRARED EYES IN THE SKY

The James Webb Space Telescope will be the largest astronomical telescope ever put into space. Spitzer, the current infrared telescope, is tiny by comparison. The size of the mirror makes the biggest difference in a telescope's light-gathering capability, and is much larger in the Webb telescope.

Mirror sizes

Spitzer
0.85 metres

Hubble
2.2 metres

Webb
6.5 metres

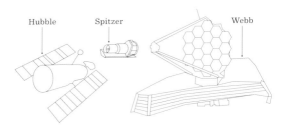

Hubble   Spitzer   Webb

Opposite: Optical telescopes of the European Southern Observatory (ESO) nestle on the slopes of the La Silla mountains in the Atacama Desert, along the Andes.

Above: The Spitzer Space Telescope was built to last a minimum of 2.5 years but it lasted for 5.5 years in orbit, tracking the Universe and reporting back to scientists.

To examine Kepler's intriguing initial findings a host of other telescopes were brought in to explore this star system, including the Spitzer Space Telescope and the European Southern Observatory in Chile. Confirming the size, orbit and density of the planet suggested that this world was not only in the habitable zone but that in the words of the scientists leading the exploration was a planet 'likely [to have] a thick gaseous envelope'. Kepler had seemingly discovered a rocky world, not too hot and not too cold, that could possibly support an atmosphere of hydrogen and helium but perhaps much more. On top of that, this was a planet around a star dim enough to allow its atmosphere to be analysed directly, and with an orbit of just 33 days repeated measurements would not take long to complete. No wonder K2-18b became the subject of intense interest among scientists hunting for alien worlds.

With such a tantalising opportunity it was time to bring in the big guns. The Hubble Space Telescope, perhaps the greatest of our orbiting galactic explorers, was instructed to point its mighty gaze towards the constellation of Leo and directly examine the spectra of starlight travelling through the atmosphere of this distant world. Hidden in the light that had passed through the envelope of gas surrounding this distant planet would be the signature of its composition. Across 1,170 trillion kilometres of space, Hubble was about to look inside the gossamer-thin atmosphere of another habitable world. With the composition of its thick atmosphere analysed by Hubble, the data was beamed back to Earth with much anticipation, and in a business that is more often than not full of failure, it didn't disappoint.

Two billion years ago, while Earth was slowly nurturing the nascent life that had begun to colonise it, a star was born in a distant part of the galaxy. A cool, red star around which at least one or two planets formed out of the swirling cloud of dust and gas that was left over from this stellar birth. The outermost of these planets, a giant rocky world, settled into an eccentric orbit around the dim star that left it skating around the edge of the habitable zone of the star system – not too hot nor too cold. And because of the substantial size of the planet, its mass could generate a powerful gravitational force that enabled it to cling on to a thick atmosphere swirling around in violent weather patterns driven by the volatile behaviour of the nearby star.

For 2 billion Earth years this world played out a life in fast forward, completing an orbit every 33 days until one day, as it made its rapid passage around its star, it pointed in the direction of our solar system 110 light years across the galaxy and created the tiniest of dips in the light of the star. This flicker registered in the eye of our most successful planet hunter to tell us to look towards this world; a world just close enough to allow our most powerful telescope to direct its gaze towards this planet and to read across many trillions of kilometres of space the code contained in the light passing through the atmosphere of this distant world. And buried deep within the data of that distant light was a remarkable signature, a breakthrough in our exploration of the galaxy we live in.

Hubble provided the first direct evidence of the existence of water vapour in the atmosphere of a planet orbiting in the habitable zone around its star. The measurement itself was far from accurate, with a wide margin of error that estimated the amount of water vapour in the atmosphere to be somewhere between 0.01 per cent and 50 per cent. Let's not forget this planet is a long, long way away and the fact that we can attempt to analyse the constituents of its atmosphere in any way at all is quite frankly astounding, but despite the wide margins the observation

THE DEW POINT
The dew point is the temperature at which air is saturated with
water vapour, the gaseous state of water. Below the dew point,
liquid water will begin to condense. A dew-point temperature
of 16 degrees Celsius or lower is comfortable for most people,
but at a higher dew point of 21 degrees C, most people feel hot
or 'sticky' because the amount of water vapour in the air slows
the evaporation of perspiration and prevents the body cooling.

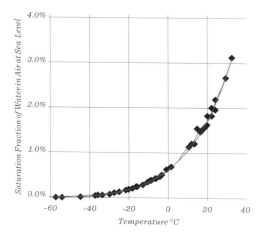

is still massively significant for two important reasons. First, it is not zero, there
is water vapour in the atmosphere of K2-18b and that is the first time we have ever
been able to say that. Secondly, even if the actual amounts of water vapour in the
atmosphere of K2-18b are at the lower end of the observation, perhaps just a few per
cent of that found in the atmosphere on Earth, it is still enough to suggest that the
conditions on this planet could be consistent with the conditions needed to create
only the second water world we know of in the Universe. A world where in its thick,
dark atmosphere clouds hang heavy with water. Holding on to this precious cargo
until the load grows so great that rains fall on this alien world, feeding the vast
oceans with an endless supply of the liquid of life. K2-18b is exciting because it's the
smallest world with an atmosphere we can analyse and we've found that its mass
and density, composition of its atmosphere and its orbit are consistent with it being
a rocky world. It's definitely a world with water, and it might be a world with oceans.

At the moment we can only dream of what might lie beneath this vision of vast
oceans. On a planet this size they could stretch to unfathomable depths, descending
for hundreds of kilometres below the surface down into a darkness deeper than
anything we have experienced on Earth. All we know is that whatever the makeup
of the planet might be, it will almost certainly have a molten rocky core, a vast
source of energy that will in places have broken through the crust and be releasing
its heat into the liquid ocean above. Creating that familiar triplet of potential – a
ready supply of energy, elements and water – all combined on a planet for perhaps as
long as 2 billion years. K2-18b offers us the first tantalising possibility that we are
looking at a world that may be a home, a planet that can support and nurture life and
that we have only just begun to explore.

Life within the clouds

K2-18b is a huge milestone. It's what we call a sub-Neptune-sized exoplanet. It's about two and a half times the size of Earth. It's almost certainly not a rocky planet, but what's so intriguing about K2-18b is that models infer that there could be some liquid water in clouds in its atmosphere, because our own Earth has life in the clouds. Bacteria get swept up from the surface and they go into cloud droplets or float around for about a week, then these bacteria, some of them, are transported over entire continents before they're dropped back down. So I love to speculate that perhaps there's some kind of life in the clouds on K2-18b; it's the first planet that we know of whose atmosphere we've observed that we can seriously speculate about.

Sara Seager, Professor of Planetary Science, Physics, Aeronautics and Astronautics, MIT

In 1936 Edwin Hubble famously described the history of astronomy as 'a history of receding horizons' and here we are finding a new horizon rapidly receding. In the galaxy that Hubble revealed to be just one of many, we are bringing into focus the distant horizons of alien worlds like K2-18b. And over the coming years we will be sending the next generation of telescopes out into the cosmos to bring those horizons ever closer.

Leading the way will be the James Webb Space Telescope, due to launch in November 2021. Webb will carry technology with far greater sensitivity than anything that has gone before. Exploring the Universe with an unparalleled sensitivity to infrared wavelengths of light, Webb will not only lead the hunt for new terrestrial exoplanet worlds, it will also have the sensitivity to unpick the chemistry of the atmospheres of distant worlds like K2-18b's in far greater detail than Hubble ever could. In doing so we may even find not just the atmospheric signature but also some clues to what may lie beneath. The Webb telescope will hopefully allow us to measure the abundance of tell-tale gases like oxygen and methane in amounts that might suggest they could only have been produced by some form of life.

K2-18b won't be the only target in Webb's sights, we already have a number of candidate exoplanets to explore further, including the TRAPPIST-1 system that has within it TRAPPIST-1e, another Earth-sized planet around a red dwarf star that we believe to be in the habitable zone. It's not impossible to imagine that by the end of this decade we may well have compelling indirect evidence of the existence of life on a number of planets across the galaxy. The first answer to the question 'are we alone?' could finally be within reach.

Opposite: Salt deposits at the Valle de Luna at San Pedro de Atacama indicates the presence of water at some point, and of life.

Right: Clouds of life-giving water hang over the tropical rainforests of Borneo.

'We humans are born explorers. We want to find other planets because we want to know if there's any life out there. And planets, rocky planets like Earth, are the place to search.'
Sara Seager, Professor of Planetary Science, Physics, Aeronautics and Astronautics, MIT

The Milky Way is an incredibly big place. We estimate that there are around 20 billion potentially Earth-like planets in our galaxy – that's rocky worlds in the habitable zone around the star, worlds that could support liquid water on the surface in our galaxy. That is 20 billion potential homes for life. Now we don't know the probability that given the right conditions life will begin on a planet, but we do have evidence from our world. What we know is that here on Earth life began pretty much as soon as it could. After the Earth had formed and cooled down, after the oceans had formed on its surface. So that might suggest that while there isn't a sense of inevitability about the origins of life, given the right conditions it might at least be reasonably probable.

So it seems that there is at least a chance that life has begun on some if not many of the 20 billion Earth-like worlds out there in our galaxy. But there are two questions about life. One is about the origins and the existence of microbes, but often when we talk about aliens what we really mean is not microbes but complex creatures, things we can speak to, civilisations. What is the probability that there will be other civilisations out there in the Milky Way? Well, the answer is again that we don't know, but there are observations that we can make, patterns we can see in the Milky Way that might allow us to make an educated guess.

Left: Scientists search for microbes and bacteria as signs of life on planets within our galaxy, the existence of inhabitable worlds.

THE SMELL
OF RAIN

LEO

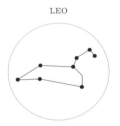

In 2015 Kepler pickled up the flicker of a planet crossing in front of a star 124 light years from Earth in the constellation of Leo. This was not by a long way the smallest of worlds that Kepler had found. It was about twice the radius and eight times the mass of our planet, putting it in the super-Earth category of exoplanet discoveries, a world where the increased gravity would dwarf the forces we see play out on our own planet. This world wasn't orbiting a sun-like star either; its parent star was a highly active and volatile red dwarf, the smallest and coolest type of main sequence star. These long-lived stars are the most numerous type of star in the galaxy, but they are not necessarily the most conducive to the creation of home. But for everything that counted against this world being a hopeful target in our search for alien life, there were also some intriguing other details in the data.

Reaching 21 million kilometres from its star and completing one orbit in just 33 days meant this world was close to its star and almost certainly tidally locked, with one face forever facing inwards. But around a red dwarf star such close proximity does not necessarily mean intense heat; in fact the data suggested that this planet, named K2-18b (because it was the eighteenth planet discovered during the 'second light-K2' part of the space telescope's mission) was potentially in the habitable zone of this cool star with a likely surface temperature hovering around zero degrees Celsius. And that was not all that made this world stand out from much of the Kepler data. A planet circling in the habitable zone of a red dwarf like this opened up the possibility of not just observing the planet, but with much less glare from the star it might be possible to explore the atmosphere of this world as well.

KEPLER'S TIMELINE

March 2009: NASA's Kepler Space Telescope launched on a three-stage Delta II rocket, from Cape Canaveral Air Force Station in Florida.

April 2009: Kepler awakens in a star-studded patch of sky in the constellations Cygnus and Lyra. Out of the 4.5 million stars, the telescope would almost continuously monitor more than 170,000 for dips in brightness as planets cross in front of their parent stars.

January 2010: Kepler's first five planetary discoveries, named Kepler-4b, -5b, -6b, -7b and -8b, are hot Jupiters – gas-giant infernos – orbiting their parent stars in just days, with surface temperatures above 1,000 degrees Celsius.

January 2011: The first solid evidence of a rocky planet comes into view with Kepler-10b, a 'lava world' orbiting so close to its star

that its star-facing side could be an ocean of molten rock. Kepler discovers hundreds of rocky planets.

15 September 2011: While Kepler-16b experiences double sunsets, it turns out to be a gas giant that might lack a solid surface.

5 December 2011: Kepler-22b is the first planet found in the 'habitable zone'. At more than twice Earth's diameter, it is among the most common planet sizes.

18 April 2013: Kepler's discovery of a new planetary system brings into view three 'super-Earth' size worlds in the habitable zones of their parent stars – Kepler-62e and -62f, and Kepler-69c (though subsequent studies have shown that Kepler-69c no longer fits this category). The discoveries prove that small planets in the habitable zone of their parent stars exist beyond our solar system.

14 May 2013: The Kepler spacecraft loses its second reaction wheel, bringing an end to science observations. The mission team searches for another way to operate the telescope in order to restart science observations.

17 April 2014: Kepler data reveals the first Earth-size planet in the habitable zone of its star. Kepler-186f orbits a cool, red dwarf star about 580 light years away. Its discovery is a step closer to finding Earth-like worlds.

May 2014: Kepler begins a new mission as 'K2' using a technique to take advantage of the pressure of sunlight to help stabilise the telescope's pointing. This requires the spacecraft to switch its field of view every three months, bringing many new patches of sky under Kepler's gaze.

23 July 2015: The Kepler team discovers a super-Earth 60 per cent larger than our

planet. Kepler-452b orbits in the habitable zone of a Sun-like star and is considered the closest found to an Earth-Sun analogue. A recent study disproves this.

21 October 2015: Kepler's extended K2 mission uncovers evidence of a small, rocky planet being torn apart as it spirals around a white dwarf star, allowing astronomers to witness the final stages of a solar system.

January 2016: Named after astronomer Tabitha Boyajian, Kepler picks up strange fluctuations in its brightness, found to be a dust cloud moving around 'Tabby's star'.

10 May 2016: The Kepler mission hauls in more than 1,200 exoplanets; many could have an Earth-like composition.

22 May 2017: Using data from Kepler's extended K2 mission, astronomers pin

down the orbital period of the outermost planet in the TRAPPIST-1 system – home to seven Earth-size planets. The data backs up the theory the planets migrated inward during the system's formation.

19 June 2017: The most comprehensive catalogue of exoplanets from Kepler's first four years of data includes 4,034 candidate planets, with 2,335 confirmed. The initial estimate of near Earth-size, habitable zone planets was 30. Since then, new data and preliminary analysis shows the actual number is possibly between 2 and 12.

14 December 2017: An eighth planet is found in the Kepler-90 system, equal to our own solar system in having the largest number of known planets. All crowd closer to their star than Earth to our sun. The discovery is made, in part, using artificial intelligence.

11 January 2018: An Australian car mechanic discovers a four-planet system with Neptune-size worlds within the data. Scientists find a fifth planet.

18 April 2018: Kepler's successor, the Transiting Exoplanet Survey Satellite, rockets into space. TESS uses the technique pioneered by Kepler – watching for dips in starlight – to find exoplanets orbiting bright, nearby stars.

May 2018: Kepler finishes six months of observing supernovae, capturing the beginning stages of these stellar explosions with unprecedented precision to resolve the mystery: What sets them off?

30 October 2018: Kepler completes its nine-year odyssey observing more than half a million stars. Kepler's discoveries have taken us one giant step further in our search for life in the galaxy.

Above: Checks on the
6.5-metre mirror (left) and
the five layers of the full-scale
sunshield (middle) before
testing in conditions that
replicate the hard vacuum
and -231 degrees Celsius
temperatures of space (right).

Right: Hopes are high for
the James Webb Space
Telescope, and a full-size
replica of this engineering
feat has toured the US.

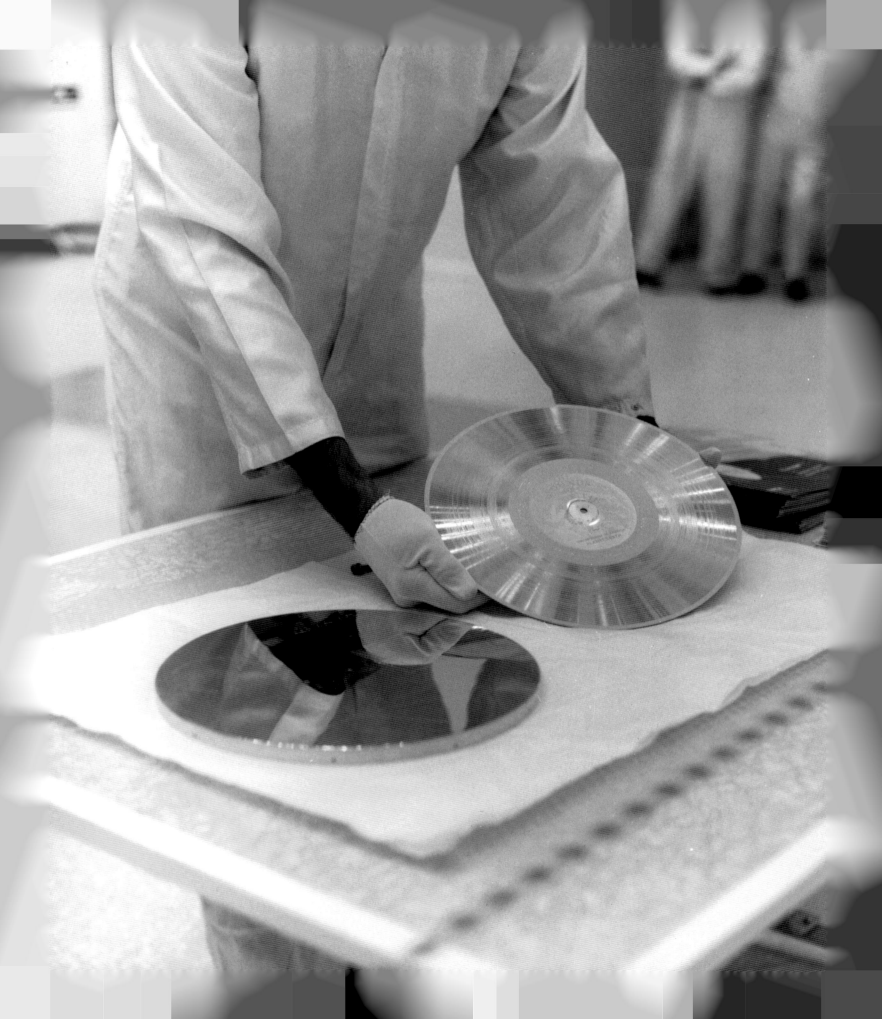

LOST WORLDS

'The spacecraft will be encountered and the record played only if there are advanced space-faring civilizations in interstellar space, but the launching of this "bottle" into the cosmic "ocean" says something very hopeful about life on this planet.'
Carl Sagan

O ur civilisation, all of the knowledge and exploration that has been amassed in little more than the past 100 years, is the product of a drama that has been playing out for a quarter of the age of the Universe. From humble microbe to a technologically advanced civilisation in 4 billion years, we are now on the cusp of finding life on other planets, reaching out across our galaxy in the hope that we will connect with another civilisation.

But for now at least we remain surrounded by silence. The messages we've sent out into the cosmos go unanswered and the telescopes that we've used to scan the skies for alien signals are empty of any form of contact. There could, of course, be endless other civilisations out there, millions of histories that are playing out across the cosmos just waiting to be discovered if we knew what to look for and where. But how likely is that really? As we sit here engulfed by the seemingly unending silence of the Universe, not yet equipped with the technological capability to listen a little further, a little deeper, a little more carefully into the darkness, we have to try to answer the question of our solitude by looking closer to home.

That's not to say that there aren't any other civilisations out there. It took 4 billion years of stability for an advanced civilisation to emerge here on Earth, and 4 billion years were also needed to arrive at a few decades of cosmic exploration. That is a vast amount of time where the Earth, the Solar System and our place in the galaxy have all been stable enough to allow life to begin and then evolve in an unbroken chain. As we've already seen, the conditions for the genesis of life – the coming together of chemical elements, energy and liquid water – may be commonplace across the galaxy and presumably the Universe beyond, but the circumstances necessary to allow that life to develop further – stability and time – may not be so abundant. As we look out to other worlds in the Milky Way it seems these two things may be a very rare commodity indeed.

We have known for a long time that the Universe is a malevolent place. Our planet has survived countless brushes with destruction. Our solar system is littered with the debris of violent interactions and filled with planets and moons that lost their chance of being a home long ago. And out there in the galaxy, the more we look the less we see a nurturing cosmos, a universe ready to cradle any emerging life.

The Gaia Telescope has revealed the Milky Way in extraordinary detail, allowing us to explore its structure, its history and its future. But as well as creating the most accurate map of our galaxy, Gaia has also given us something else, a new perspective on our solitude.

As it has catalogued and mapped over a billion stars on its five-year mission, Gaia has revealed that our own solar system with its solitary star at the centre of a host of orbiting planets is far less common than we had ever imagined. It seems that the majority of stars in the Milky Way are not alone, they are twins or triplets, multiple star systems orbiting around each other in a complex gravitational dance. The existence of multiple star systems has been known about for a long time, with the discovery of the first true binary star system by William Herschel in the 1700s, but until very recently we had little idea how common they are. Now we have a huge amount of high-precision data, including from the Gaia Space Telescope, and what this tells us is that the vast majority of star systems are not singular (around 80 per cent of giant stars are not alone) and when it comes to main-sequence stars like our sun, over 50 per cent are in multiple star systems. That means if we could

Opposite and left: Two golden phonograph records were aboard both Voyager spacecraft in 1977 as a sort of time capsule for any intelligent extraterrestrial life form who may find them. The records contain sounds and images selected to portray the diversity of life and culture on Earth.

drop down onto a random planet anywhere in the Milky Way, it's far more likely we would witness a multiple sunrise than not. But as well as the undoubted beauty it must bring to these worlds, it also means that the majority of planets face a far more perilous existence than we had thought.

In September 2020 an Earth-sized exoplanet was discovered by a group of international scientists working on the Optical Gravitational Lensing Experiment (OGLE). Such an announcement is not unusual in the age of Kepler and all the other planet-hunting telescopes that we have staring out into the Milky Way, but this discovery was different. This planet named OGLE-2016-BLG-1928 (we'll call it OGLE from now on) was an Earth-like world but with something fundamental missing, because OGLE was a planet without a parent star. Finding a rogue planet like this, gravitationally detached from any star and so free-floating and lost in the darkness of space, is far from easy. Our normal exoplanet-detecting techniques that require the distortion of light from a parent star as the planet orbits is not possible when there is no starlight to be distorted.

Instead, planet hunters looking for rogue planets need to use a technique called microlensing, which requires a planet to pass between us here on Earth and a distant light source, a star somewhere in the background. Emerging from Einstein's Theory of General Relativity, the idea is that a massive object passing between us will bend and magnify (lens) the light of the star and in doing so make the planet detectable from Earth. It's literally like looking for a needle in a cosmological haystack and makes normal planet hunting look easy. It's estimated that if we observed one

Rooting out the rogue planets

A rogue planet is a planet without a star, one that was probably cast out of its planetary system, which is now a freely floating body drifting through the galaxy, untethered and homeless. We think there are a lot of rogue planets out there – at least one for every star. After a star is born, the leftover discs of gas and dust create more stars, and as the planets settle into their orbits and become stable, some planets are ejected. The study of these rogue planets is important to help us to understand the life cycles of planetary systems, from birth to death.

Sara Seager, Professor of Planetary Science, Physics, Aeronautics and Astronautics, MIT

source star waiting for a microlensing event we would on average have to wait a million years before we would see one planet cross its path. That's why astronomers hunting for objects using microlensing have to survey hundreds of millions of stars to give themselves a chance, and in the case of the OGLE sky survey have been doing so for nearly 30 years.

The result is that we are now starting to detect these lost worlds and OGLE-2016-BLG-1928 is potentially the smallest and most Earth-like rogue planet we have detected so far, but as far as we can tell OGLE is anything but unique. It's estimated that there may be over a hundred billion rogue planets in the Milky Way, making it perhaps the most common type of planet in our galaxy. To understand why rogue planets like OGLE may be so commonplace we need to understand their origin, where these lost worlds come from.

As far as we know, planets can only form around the gravitational influence of a parent star, so every planet in the galaxy – be it rogue or not – has to begin its life in a star system. With the data we have we can't be certain of OGLE's origins, but the evidence suggests it would have most likely begun life in the most common type of star system in the galaxy – a twin or multiple star. Dawn on this planet would have seen at least two suns rise in the morning sky, but for all the beauty of a binary star system there is chaos lurking in such systems. For OGLE, a rocky world with an Earth-like mass, its orbit around this star system would have exposed it to a gravitational tug of war as the stars fought to exert their own gravitational dominance on the planet. A world caught in a stellar tug of war.

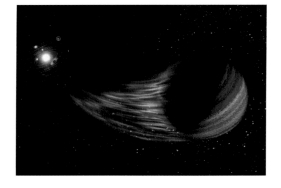

Above: An artist's image of a rogue planet that has been ejected from its home planetary system.

Right: Illustration of OGLE-2005-BLG-390Lb, located some 3,300 light years from Earth. Although five times our planet's mass, it is believed to have a rocky core and a thin atmosphere.

'Orbiting a binary star system, our orbit may become unstable. In the same way, an asteroid flying through the Solar System is cast way out before it comes back. You can imagine something like that casting a planet out into space and leading to what we call rogue planets.'
Phil Muirhead, Astrophysicist, Boston University

Planets in these very common types of star systems live an inherently precarious existence, because they are subjected to the gravitational pull of two stars. Even in single-star systems the weak gravitational interactions between the planets can change their orbits. Now in a double-star system the planets are not only subjected to the gravitational pulls of each other, they are subjected to the even stronger gravitational pull of another star. So even if a planet gets into a stable orbit, it's very likely that it won't stay in that orbit for long. This means that in double-star systems the line between order and chaos is very thin indeed.

OGLE would almost certainly have been just one of a number of planets orbiting the two stars and over time the stability of this planetary system would have inevitably been disrupted. Just as we suspect happened in our own solar system, outer-lying giant planets like Jupiter can often be flung inwards by gravitational

Below: The Whirlpool galaxy
is actually a pair of galaxies
also known as Messier 51
and NGC 5194. The winding
arms of the spiral are long
lanes of stars and gas laced
with dust, which act as star-
formation factories.

perturbations and that's far more likely to happen in a multiple star system than with a single star like our own. Not only is this orbital disruption more likely but the consequences are potentially more profound as well. In the case of a rogue planet like OGLE, the trajectory of an inbound planet could have easily crossed OGLE's path, delivering a decisive blow to the planet that would have knocked it not just out of its orbit but out of the entire star system itself. Such cosmic pinball sounds unlikely, but this is our best current explanation for how rogue worlds like OGLE came to be, and as far as we know a world without starlight is a world without life. Today, far from the warmth of its parental star, any liquid that OGLE might once have had has long ago frozen solid, any atmosphere that once protected the rock surface from the harsh cosmic rays is also long gone. This and its band of rogue brothers are all sterile worlds, haunted by memories of a youth spent bathing in the potential of a mother star.

The story of these rogue worlds illustrates the true nature of the galaxy we live in. A chaotic environment where planets are more likely than not to experience monumental change. The odds seem stacked against the stability that we think complex life needs to progress to a civilisation. There may be endless habitable worlds out there but very few that dodge the events that conspire to stop life in its tracks. Yet amongst the galactic trend for chaos and change, one planet's story stands in stark contrast.

Our planet appears to have largely escaped the violence, the chaos and the constant change that seems to characterise a galaxy like the Milky Way. Yes, there's been the odd mass extinction, but there's an unbroken chain of life here on Earth stretching back 4 billion years. And if that's what you need to go from the origin of life to a civilisation, although there might be billions of worlds out there where life began there may be very few civilisations.

But that's just an opinion, an educated guess. And given the profound nature of the question, no matter how educated the guess it would be ridiculous to stop looking, both inside our galaxy and beyond.

In September 2020 astronomers using NASA's Chandra X-ray Telescope detected a tell-tale flickering light not in our galaxy but in the Whirlpool galaxy, M51, 26 million light years from Earth. It's not fully confirmed yet, but M51-ULS-1 is the first planetary candidate to be detected in a galaxy beyond our own. What we think may be a giant gas world, just a little smaller than Saturn and orbiting in an incredibly bright binary star system quite literally in a galaxy far, far away.

The potential discovery of a planet orbiting around a star in another galaxy is something many scientists never thought they'd see. And it opens up the intriguing possibility that we might be able to explore not only the question are we alone in the galaxy but are we alone in the Universe? Now the answer to 'are we alone?' may be answered far into the future, and indeed might never be answered, but the question is profound because answering it would teach us much more about what it means to be human.

And we become a little bit more human with every world we explore, because that ability to lay the foundations, to explore questions to which we may never receive answers in our lifetime, questions for our children or our grandchildren to answer, is a fundamental part of what it means to be human. It's a fundamental part of what makes us so special on this little world, looking up at the stars: are we alone or not?

STARS

'Stars, hide your fires;
Let not light see my black and deep desires.'
William Shakespeare, Macbeth

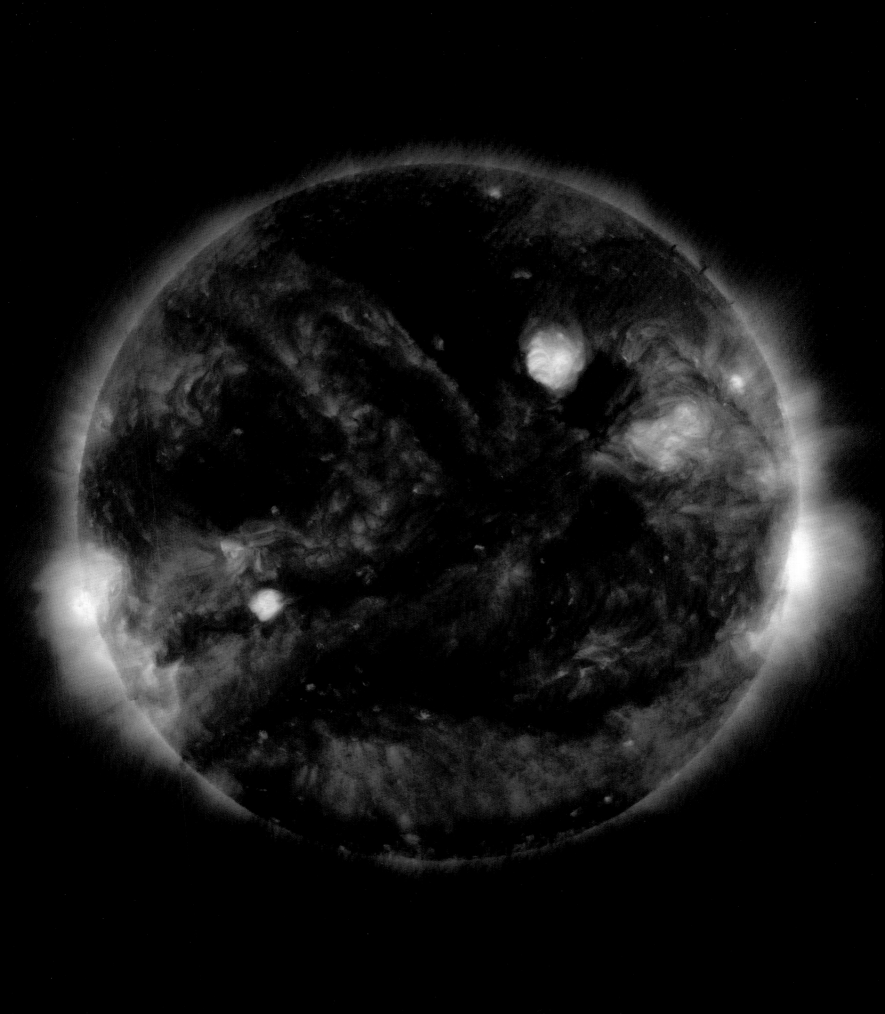

THE STARS

There was a time before starlight, a time before the heavens were bathed in the glow of billions upon billions of stars. This was a dark universe, filled with huge clouds of swirling gas, devoid of any visible illumination; it was frozen, stunted in its infancy and locked in a seemingly eternal stasis. Without stars there was no light, no creation, no engine to propel forward the embryonic Universe. It was a universe made of the simplest of elements, strung together on the darkest of webs; a cosmos without complexity; no stars to make stardust, no stardust to make rocks, no rocks to make planets, no planets to make worlds. And in a universe without worlds there can be no habitats – no such thing as life.

And yet today we look up from our living, breathing planet and see a very different story. Bathed by the light of a single star, we wait for our sun to set before the abundance of our universe can be revealed. Above us the night sky is lit by countless stars, just a few thousand of which are visible with the naked eye from any point on Earth. Among those stars the seven other planets of the Solar System appear and disappear as they dance across our darkening skies.

We take all this for granted – the rising and the setting of the Sun each day – and we have historically and confidently configured our days, our nights and our calendars by its motion, and even in these days of technological advances we cannot escape the basic premise that we are connected to the cosmos, because, fundamentally, we all come from stars.

Staring up at the night sky is just the beginning, the limit of where we can travel with the human eye alone. Beyond our gaze there is a galaxy more abundant than we can imagine, with more than 100 billion stars swirling together in their spiral form. Hidden behind their glare are billions of planets, worlds of almost endless variation, some giant and formed from gas, others rocky and perhaps covered in the oceans that are so critical for life. If we could peer farther into the darkness around these planets we would see more worlds appear – we can only estimate how many there are, the myriad moons of the Universe that we know must exist in such mind-boggling numbers.

Our Milky Way is just one galaxy among the 100 to 200 billion galaxies within reach of our largest telescopes that fill the Universe with endless, unimaginable potential. All of it, a universe full of possibilities (every star, every planet, every moon, every rock, every tiny speck of dust) would not exist without the moment that took us from darkness to light.

In this chapter we will explore the story of that momentous transformation, we will travel back billions of years to see how those very first stars came into being and discover how they in turn created the conditions for everything that followed. This is the story of our stellar family tree. A tree that has, through just a handful of generations, transformed our universe into a cosmos of starlight, and in the process created a one-star system, within which the chemistry on one planet has created a living world. Perhaps the only living world that has reached out across millions of kilometres of space and tapped into the endless supply of energy from its nearest star. Fuelling not just life but a lifeform that has built a civilisation capable of exploring the Universe from which it came and the destiny that universe will ultimately bring.

'All of the elements in stars are so important to us because we come from the stars. Most everything that we're familiar with today – the building blocks of the trees, the homes we live in, the clothes we wear, the elements in our body – they were created, they were forged within the very heart of stars billions of years ago. So stars, not just the Sun, but stars in general, are responsible for our existence.'
Nia Imara, Astrophysicist, University of California

Opposite: The night sky is lit with billions upon billions of stars, and the Milky Way stretches above us, offering a universe of possibilities.

INTO THE LIGHT

Our story begins relatively close to home, just 150 million kilometres from our planet. Our nearest star, the Sun, lies at the centre of our solar system. At 4,370,005.6 kilometres in circumference and containing 99.8 per cent of all mass in our solar system, the scale and violence of our stellar neighbour has been deified by ancient cultures and inspired myriad myths and legends since humans first looked up and wondered what this life force could be and where it had come from.

At first with our eyes, then with the help of ever-more-powerful telescopes, then with the help of the Hubble Space Telescope, and now with a fleet of intrepid hi-tech explorers, we have been able to build our understanding of what lies within the realm of this near-perfect sphere. But it's not only been physical observation that has led us to the solar truths, it's also because of theoretical exploration, the journey of the human mind to reach out and understand the mechanism by which the Sun functions. Through this combination of observational and theoretical study, we have been able to piece together the true origin of the Sun's power that rains down on the surface of our planet, and to rewind the life story of this hot ball of glowing gases, this giver of life.

We begin at the turn of the last century, in the first years of the 1900s, with the revolution in our understanding of the physical world that began with the work of Ernest Rutherford in his Manchester laboratory. As our understanding of the atom grew and our knowledge of the fundamental basics of the physical world expanded, so our ability to transpose this knowledge onto the grand structures of the Universe, structures like the Sun, allowed us to begin a parallel journey of discovery.

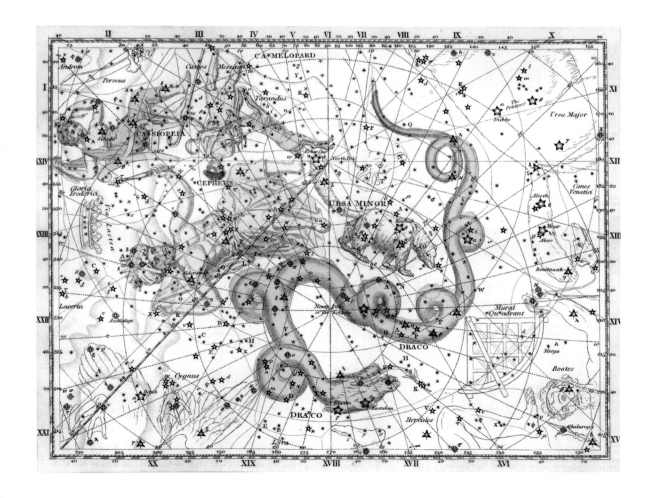

Left: Through 30 maps in his *Celestial Atlas* of 1822, Alexander Jamieson displayed the heavens as then visible to the naked eye.

CALENDARIO AZTECA O PIEDRA DEL SOL.
EN EL MES DE DICIEMBRE DEL AÑO DE 1790
AL PRACTICARSE LA NIVELACION PARA EL NUEVO
EMPEDRADO DE LA PLAZA MAYOR DE ESTA CAPITAL
FUE DESCUBIERTO ESTE MONOLITO Y COLOCADO
DESPUES AL PIE DE LA TORRE OCCIDENTAL DE LA
CATEDRAL POR EL LADO QUE VE AL PONIENTE
DE CUYO LUGAR SE TRASLADO A ESTE MUSEO
NACIONAL EN AGOSTO DE 1885.

UNLOCKING THE ATOM

Rutherford proved the existence of the atomic nucleus by updating Thomson's 'plum pudding' model, which had assumed uniform positivity with negative electrons scattered throughout.

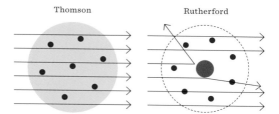

Modern science had puzzled over the source of the Sun's heat for centuries, so replacing long-held mythology with an evidence-based explanation for its extraordinary power was not easy. Was the Sun simply born incredibly hot and began to gradually cool as it shared its heat with the Solar System? Or was there some kind of mechanism that was generating heat from within, gravitational contraction perhaps, or was the heat produced by meteorites falling into the star?

Many theories came and went through the frenzy of nineteenth-century science, but it was Rutherford who was the first to propose a theory for the power of the Sun. Built on his laboratory work, he proposed that the Sun's heat emanated from the disintegration of the atomic nucleus, and the resultant release of this heat was behind the power of the Sun. The theory that a star is radioactive in nature would not ultimately prove to be correct, but it pointed us in the right direction and marked the beginning of a theoretical model of the Sun based on atomic structure that would be built on by other scientists over the coming decades.

It was Albert Einstein who picked up the baton next, with his principle of mass – energy equivalence – the extraordinary idea that anything having mass has an equivalent amount of energy, and vice versa, and that you can convert from one to the other using a constant – 'c^2' – which is a number whose value is equal to the square of the speed of light. Shared with the world for the first time on 21 November 1905, the expression of this fundamental principle in the equation $E=mc^2$ created

Above: New Zealand-born British scientist Ernest Rutherford was the greatest experimental physicist of his day, winning the Nobel Prize in Chemistry in 1908.

Right: Ernest Rutherford (right) and Hans Geiger together devised a method of detecting alpha particles and counting the number emitted from radium.

'A star is drawing on some vast reservoir of energy by means unknown to us. This reservoir can scarcely be other than the sub-atomic energy which, it is known, exists abundantly in all matter; we sometimes dream that man will one day learn how to release it and use it for his service.'
Arthur Eddington

the foundation upon which nuclear physics would flourish. It also allowed physicist and populariser of science, Arthur Eddington, to describe nuclear fusion as the fundamental mechanism of energy production within a star such as our sun. With his 1925 publication, *The Internal Constitution of the Stars*, Eddington travelled across 150 million kilometres of space and enabled us for the first time to peer inside the inner workings of the Sun, revealing that it was in fact the simplest of all the elements, hydrogen, that was the source of almost all of the Sun's energy.

Following the principles laid down by Einstein, Eddington speculated that under the immense pressures generated by the Sun's vast mass, hydrogen atoms could be forced together, fused in a moment of cosmic alchemy – hydrogen into helium atoms – which would liberate enormous amounts of energy. Before we had reached an understanding of nuclear fusion, thermonuclear energy or even the fact that stars are largely composed of hydrogen, Eddington had taken the principle of mass–energy equivalence and surmised that four hydrogen atoms could combine to create a single helium atom. He further concluded that in the process an extraordinary amount of energy was released from the resultant net change in mass, raising the interior temperature of a star like our own to millions of degrees. Eddington theorised a star need only be 5 per cent hydrogen to explain the generation of energy. This was the beginning of our understanding of the internal mechanism of a star and the first step in understanding the life story of the stars.

Above: Eddington's dream is gradually coming true thanks to research on hydrogen fusion reactors. Leading the way is a device called JET, the Joint European Torus.

Right: The work of Arthur Eddington (right) and Albert Einstein laid the foundation for our understanding of what powers the stars.

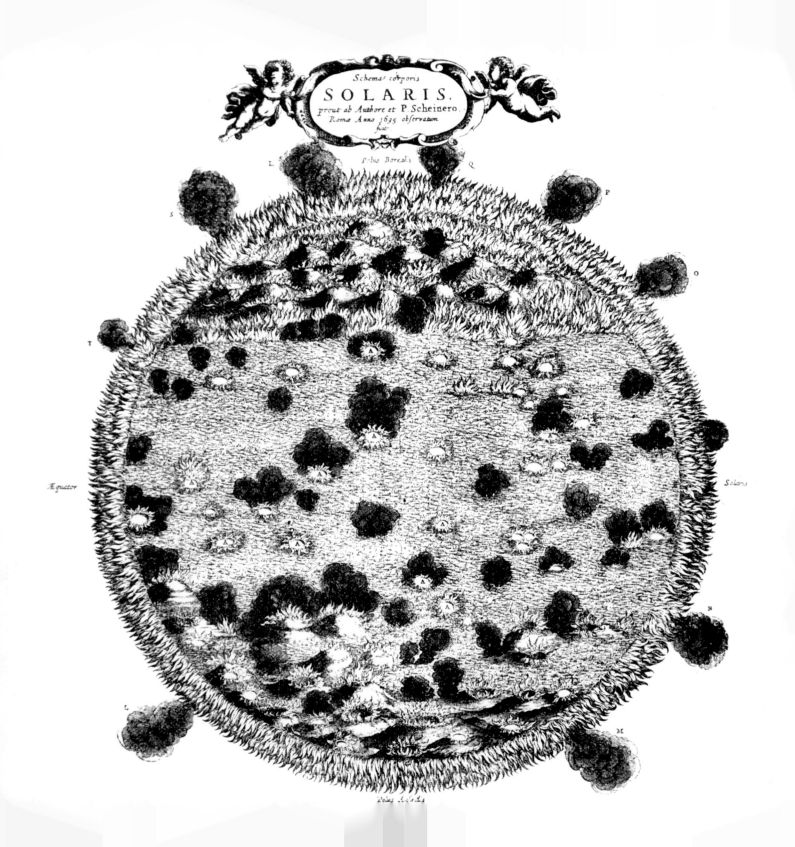

STELLAR ELEMENTS
Hydrogen, helium and lithium were the first elements
in the Universe. Their atomic structures below are
in the modern atomic model, which accounts for the
quantum orbits of electrons.

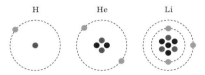

H   He   Li

● Protons
● Neutrons
● Electrons

It would be another five years and require 'the most brilliant PhD thesis ever written in astronomy' to build on Eddington's hypothesis and reveal that far from being a minor component of the Sun, hydrogen was in fact the most abundant element, not just in a star but in the Universe. The 'brilliant thesis' was written by the British-born astronomer Cecilia Payne-Gaposchkin, who, as an undergraduate at Cambridge University in 1919, had been inspired to study astronomy after attending a lecture by Arthur Eddington. Encouraged by Eddington to follow her academic passions, Payne decided there were more opportunities for a female astronomer in the United States than Britain, so she took up a fellowship to study at Radcliffe College, which was at that time the only college at Harvard that accepted women.

Working in the Harvard College Observatory, Payne used spectral analysis to calculate the composition of not just the Sun but many other stars, from which she concluded that they were composed of primarily hydrogen. (We now know that stars are around 74 per cent hydrogen by weight.) At the time of Payne's publication, the long-held consensus view was that the elemental make-up of the Sun was identical to that of the Earth, so when her dissertation detailing a theory of hydrogen abundance was sent for peer review it was met with immediate derision. The United States, it seemed, was not as progressive as Payne had hoped in accepting a female view of the Universe, and she was ultimately pressured by her reviewers, particularly Henry Norris Russell (of Hertzsprung-Russell diagram fame), into describing her groundbreaking calculations and results as 'spurious'. It would take another four years for Payne's work to get at least some of the credit it deserved, when Russell published similar findings that he obtained using a different technique. Although Russell acknowledged the pioneering work of Payne in his publication, this did not prevent the discovery of a star-filled universe fuelled by nothing more than hydrogen remaining an explanation that would be firmly locked to his name. No matter how bright the star, it seems Payne was yet another female scientific explorer to get lost in the shadows of the male-dominated academia of the time.

'No idea should be suppressed ...
And it applies to ideas that look
like nonsense. We must not forget
that some of the best ideas seemed
like nonsense at first. The truth
will prevail in the end. Nonsense
will fall of its own weight, by a sort
of intellectual law of gravitation.
If we bat it about, we shall only keep
an error in the air a little longer.
And a new truth will go into orbit.'
Cecilia Payne-Gaposchkin

Opposite: Physicist and astronomer Christoph Scheiner's drawing of how he perceived the surface of the Sun in the 17th century.

Right: Despite her pioneering work, Cecilia Payne's groundbreaking contributions were first derided by male academics.

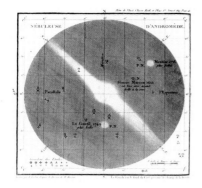

A glimpse beyond Andromeda

In 964, the Persian astronomer Abd al-Rahman al-Sufi described the Andromeda Galaxy as a 'nebulous smear' in his *Book of Fixed Stars*.

Since those ancient days, Andromeda has fascinated star gazers and scientists alike. Star charts labelled it 'the Little Cloud', then, in 1745, Pierre Louis Maupertuis theorised that the blurry spot in the skies was an island universe, before Charles Messier catalogued Andromeda as object M31 in 1764. In 1785, the astronomer William Herschel concluded Andromeda was the nearest of all the great nebulae and incorrectly estimated it to be no more than 2,000 times the distance of Sirius, or roughly 18,000 ly (5.5 kpc). In 1850, William Parsons, 3rd Earl of Rosse, provided us with the first drawing of Andromeda's spiral structure.

In 1885, a supernova (known as S Andromedae) was seen in Andromeda, the first observed in that galaxy. At the time Andromeda was considered a nearby object, so the cause was thought to be a much less luminous and unrelated event called a nova, and so it was named 'Nova 1885'.

In 1888, Isaac Roberts took one of the first photographs of Andromeda, mistaking it and similar 'spiral nebulae' as star systems being formed.

Although visible to the naked eye, Andromeda has revealed more to us about the history of our universe through the work of scientists with greater technology to hand – notably the Hubble Space Telescope. In recent years we have been offered a glimpse into the Andromeda Galaxy but also taken further, into the galaxies that lie beyond.

For the next 20 years we lived with the knowledge that we were bathing in the light of a hydrogen-fuelled sun by day and a sky filled with thousands of hydrogen-fuelled stars by night. Over this period of time our window on the Universe transformed, we took our first glimpse at a universe filled not just with one galaxy – an island universe – but full of islands of stars, galaxies like the great spiral in Andromeda, filled with more suns than we could ever have imagined. It was also during this time that we began to grapple with the idea of a universe with a beginning, middle and end. An expanding universe that could be wound back to a time before time, a day without a yesterday, a moment when the infinitesimally small became the beginning of the infinitesimally large. We slowly began to accept that we were part of a story that began with a big bang and led to a universe full of billions of galaxies containing billions upon billions of stars. But throughout the 1930s and 40s, the story of the Universe was one without a plot, without a clear sense of how the lead characters shaped and drove the destiny of it all.

That part of the story would finally be filled in in the mid-1950s with the publication of one of the most profound insights in the history of science, one that would revolutionise our perception of our sun, our planet and ourselves – revealing for the first time what we are made of and where we all come from.

Known as the B2FH paper after the four authors – Margaret Burbidge, her husband Geoffrey Burbidge, William Fowler and Fred Hoyle – this was a paper that built on the earlier hypothesis of Hoyle around the concept of stellar nucleosynthesis: the idea being that elements, the very stuff of matter, are made within stars. Many theories had been proposed about the origin of the Universe's matter over the previous few decades, including the idea that all matter was created at the moment of the Big Bang. But in a series of papers published in 1946 and 1954, Hoyle proposed that only the very lightest three elements – hydrogen, helium and lithium – were created at the moment of the Big Bang. The rest, he suggested, were the products of stars.

The culmination of this work came with the publication of the B2FH paper in 1957. On face value, this was a review paper, as much of the work contained within it had already been published, but its real value came from the combining of three different perspectives on the function of stars – Hoyle's theoretical work on stellar nucleosynthesis, the observational work of the Burbidges on the relative abundances of elements in the Universe, and Fowler's work as an experimental nuclear physicist on the nuclear reactions that were going on inside stars. However, now brought together for the first time in this paper, the result was a far more compelling and utterly revolutionary telling of the story of the Universe.

Stars do much more than just fuse hydrogen: they are the factories of the Universe, producing from the simplest of ingredients (hydrogen, helium and a little lithium) all the other elements that make up the structures of everything – you, me, everything you can see and touch, everything in nature, everything on our planet. Stars turn just one or two ingredients into the full array of the elements of the periodic table. From now on we would all be living with the knowledge that we are made of stardust. We are the product of generations of suns that have lived and died.

Opposite: This observation of Messier 31 shows blue regions of young, hot, high-mass stars where star formation is occurring; the orange-white 'bulge' is made up of old, cooler stars formed long ago.

Above: Sketch of the Andromeda 'nebula' (M31) by Charles Messier, showing Andromeda's two companion galaxies M32 (bottom) and M110 (top). Originally published in 1807.

OUR SUN

Photosphere (visible layer)
Temperature: 6,000 degrees Celsius
Density: $2 \times 10\text{-}9$ g/cm^3 (.00001% the density of air)
Thickness: 400 kilometres

Convective Zone
Temperature: 2 million to 6,000 degrees Celsius
Density: $2 \times 10\text{-}7$ g/cm^3 (.001% the density of air)
Thickness: 182,000 kilometres

Radiative Zone
Temperature: 2 million degrees Celsius
Density: From 20 g/cm^3 (the density of gold)
to 0.2 g/cm^3 (less dense than water)
Thickness: 375,000 kilometres

Solar Core
Temperature: 15 million degrees Celsius
Density: 150 g/cm^3 (more than 10 times the density of lead)
Diameter: 277,000 kilometres

Chromosphere
Temperature: 6,000 to 20,000 degrees Celsius
Density: $2 \times 10\text{-}12$ g/cm^3
Thickness: 1,700 kilometres

Transition Zone
Temperature: 22,000 to 1 million degrees Celsius
Density: $2 \times 10\text{-}13$ g/cm^3
Thickness: 100 kilometres

Corona (outer atmosphere)
Temperature: 1 to 3 million degrees Celsius
Density: $2 \times 10\text{-}13$ g/cm^3
Thickness: 29 million kilometres

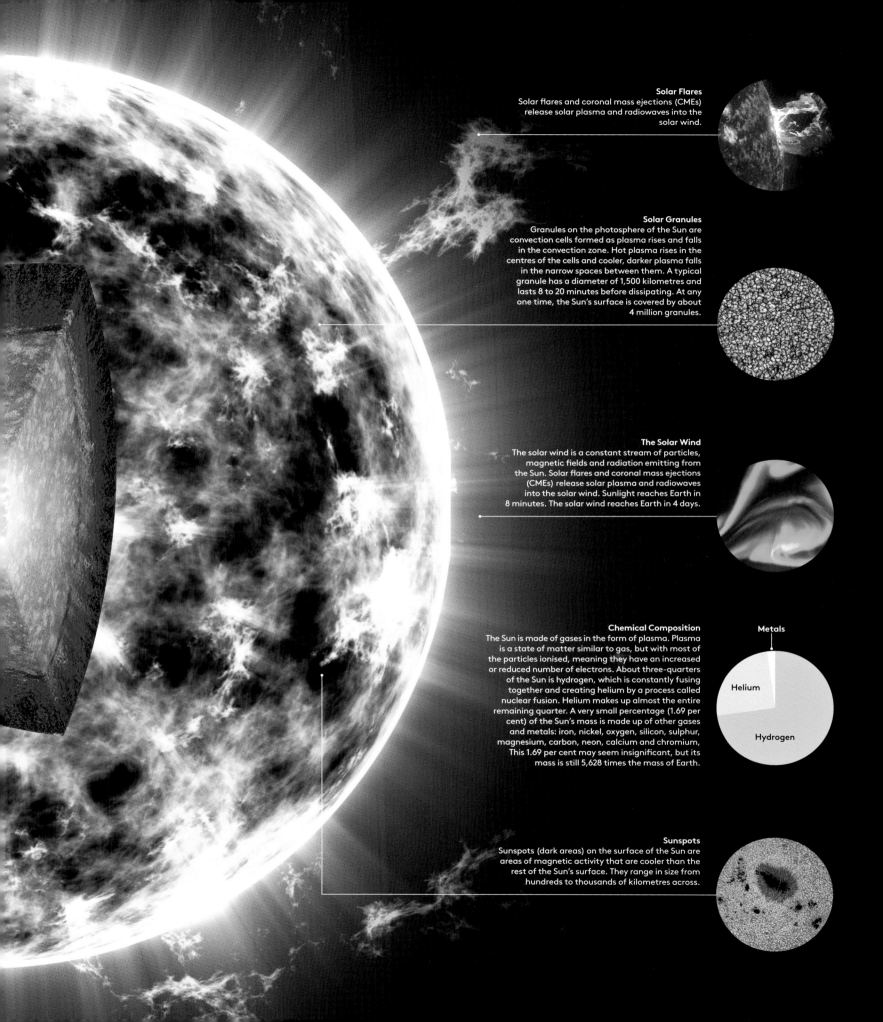

Solar Flares
Solar flares and coronal mass ejections (CMEs) release solar plasma and radiowaves into the solar wind.

Solar Granules
Granules on the photosphere of the Sun are convection cells formed as plasma rises and falls in the convection zone. Hot plasma rises in the centres of the cells and cooler, darker plasma falls in the narrow spaces between them. A typical granule has a diameter of 1,500 kilometres and lasts 8 to 20 minutes before dissipating. At any one time, the Sun's surface is covered by about 4 million granules.

The Solar Wind
The solar wind is a constant stream of particles, magnetic fields and radiation emitting from the Sun. Solar flares and coronal mass ejections (CMEs) release solar plasma and radiowaves into the solar wind. Sunlight reaches Earth in 8 minutes. The solar wind reaches Earth in 4 days.

Chemical Composition
The Sun is made of gases in the form of plasma. Plasma is a state of matter similar to gas, but with most of the particles ionised, meaning they have an increased or reduced number of electrons. About three-quarters of the Sun is hydrogen, which is constantly fusing together and creating helium by a process called nuclear fusion. Helium makes up almost the entire remaining quarter. A very small percentage (1.69 per cent) of the Sun's mass is made up of other gases and metals: iron, nickel, oxygen, silicon, sulphur, magnesium, carbon, neon, calcium and chromium, This 1.69 per cent may seem insignificant, but its mass is still 5,628 times the mass of Earth.

Metals

Helium

Hydrogen

Sunspots
Sunspots (dark areas) on the surface of the Sun are areas of magnetic activity that are cooler than the rest of the Sun's surface. They range in size from hundreds to thousands of kilometres across.

GENERATIONS

Our star, our sun, a source of unimaginable power 1.39 million kilometres in diameter, over 100 times larger than the Earth and, as we now know, driven almost entirely by the fusion of hydrogen into helium. Look up on a sunlit day and you are seeing our star transform around 600 million tonnes of hydrogen into helium every second, in a process encapsulated by Einstein's equation $E = mc^2$. In just one minute, 250 million tonnes of that matter is converted into the energy that radiates across the Solar System, shaping everything it touches. Here on Earth, that power manifests itself through the living planet that we see all around us, an unbroken chain of energy that flows directly from the Sun through almost every living thing on Earth.

Humans have looked up at the Sun for thousands of years and seen it as eternal, a neverending source of light, heat and life, but like every star, our sun has a life story. We bathe in the light of a middle-aged star that has been burning for almost 5 billion years, but contained within its make-up is a deeper secret to its history, to the story of its birth, and with that the history of the whole Universe. Through learning about our sun we can really understand what is possible in other stars – how the energy flows within it, where the wind has come off the star, and how much of it there is, as well as gain valuable insight into its life cycles.

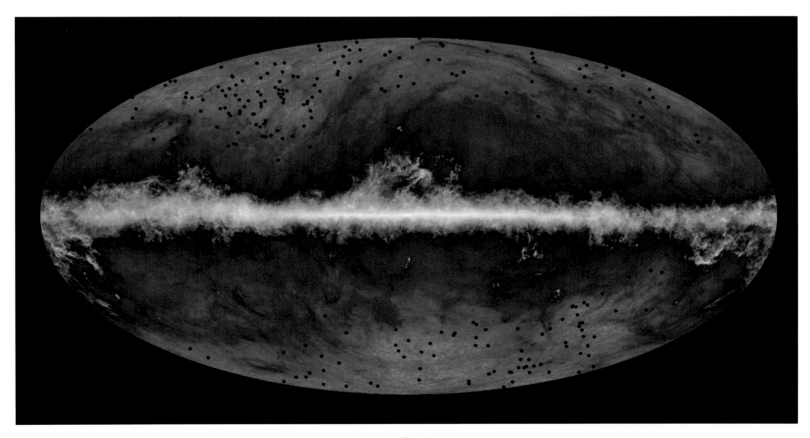

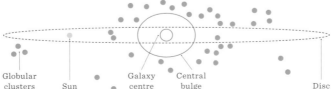

Globular clusters   Sun   Galaxy centre   Central bulge   Disc

Above: Captured by ESA's Planck, black dots indicate globular clusters across the entire sky; the central line reveals dust in the Milky Way.

A search for the first stars
Between 2012 and 2017, a team of European
researchers, led by Rachana Bhatawdekar of the
European Space Agency, set out to study the first
generation of stars in the early Universe, the
Population III stars.

　Bhatawdekar and her team probed the early
Universe from about 500 million to 1 billion years
after the Big Bang by studying the cluster MACS
J0416 and its parallel field with the Hubble Space
Telescope (pictured above). Although they were
unable to find any evidence of these stars, in the
process they revealed the deepest observations
ever made of these galaxy clusters and the galaxies
located behind them. The study also discovered
galaxies with lower masses than ever previously
observed with the Hubble Space Telescope, at a
distance corresponding to when the Universe was
less than 1 billion years old. So without any evidence
of stellar populations and low-mass galaxies, it has
been deduced that, for now, these galaxies are the
most likely candidates for the formation of the first
stars and galaxies.

Above: Globular clusters are
spherical collections of stars,
tightly bound by gravity.
There are around 150 such
clusters known in the Milky
Way, containing some of
the oldest stars in the galaxy.

At this stage of its life our sun is made up almost entirely of hydrogen (73.8 per
cent) and helium (24.8 per cent), while the remaining 1.3 per cent of mass consists
of trace amounts of heavier elements, elements that in astronomical terms are
known as metals (in astronomy, as a shorthand, all elements heavier than hydrogen
and helium are known as metals). In our sun the most abundant of these metals is
oxygen (1 per cent), followed by carbon (0.3 per cent), neon (0.2 per cent) and iron
(0.2 per cent). The presence of these heavier elements, although trivial in terms of
the composition of the Sun, is actually the clue that allows us to understand the
deeper history of not just our star but all the stars in the Universe.

In those tiny traces of metal is the story of our sun's inheritance; its position
in the family tree of stars is written in those heavy elements just as the story of our
generations is written in our genes. They tell us that the Sun is a relatively young
star in the history of the Universe, one that is rich in elements that could not have
existed in the early Universe, and so possibly a second- or third-generation star.
This is known as a Population I star, which are believed to have formed between
about 1 million and 10 billion years ago, and which are rich in the metals that must
be the product of an earlier generation of stellar lives. Older than these stars are
the Population II stars (formed between about 10 billion and 13 billion years ago),
which have relatively little metal content. This means that the gas from which these
stars formed could only have existed in the early Universe before generations of stars
churned out the richer mix of ingredients that we see in younger stars.

This concept of stellar populations was first suggested in the early 1950s by
German–American astronomer Walter Baade. Baade had been working at the
Mount Wilson Observatory in California throughout the Second World War, taking
advantage of the improved observing conditions provided by the reduced light
pollution of wartime blackouts to resolve stars towards the centre of the Andromeda
Galaxy for the first time. It was these observations that led him to define the two
generations of stars. He observed that the younger Population I stars tend to be found
within the spiral arms of a galaxy like ours, whereas the older Population II stars
tend to be found near the centre of a galaxy in an area known as the galactic bulge.

Baade's observations left the Universe with just two generations of stars, until in
1978 another line was added to the stellar family tree – a third generation made up
of the oldest of all stars in the Universe (dating back to the first few hundred million
years after the Big Bang), known, unsurprisingly, as Population III stars. Massive,
bright and hot, these stars were the first to light up the Universe and so would have
been unerringly pure with no metals at all within their make up. Formed from just
the hydrogen and helium that existed in the Universe after the Big Bang, it was
these stars that would act as the earliest factories, creating the first of the heavy
elements that would go on to build all the future generations of stars and planets.
Unlike the other two generations, Population I and Population II, that we have
observed in the night sky, these oldest of stars only exist in the minds of astronomers,
a hypothetical ancestor that we are almost certain lit up the Universe but that we
can only observe through the imagination of theoretical cosmology. These stars
take us back to a time not much more than 100 million years after the Big Bang, a
time when in an instant the Universe was transformed – when with the birth of the
very first stars it was suddenly lit up with starlight and all the potential that that
brings; starlight that emerged from the deepest darkness.

WHAT MAKES A UNIVERSE?

Rana Ezzeddine, Astronomer, University of Florida

The way we can think about the composition of the Universe is to look at it over three different points or different times in its life. So if we look way back at the Big Bang, we know that the Universe was only created a very few seconds after that moment, made only from hydrogen, helium and a little bit of lithium. If we look at about 10 million years after that, when the very first stars were formed, this is what they were formed from. They were just these elements: hydrogen, helium and lithium.

A second-generation star is formed from the gas and the material that was left over from the very first stars, the first generation. This means that whatever was found and forged inside the first stars at the end of their lives, was ejected via supernova explosions into the surrounding gas. Thus the second generation of stars bear direct evidence of the first stars, because if we can detect and study the chemical compositions of the second generation of stars, we can understand in what kind of environments the first stars existed, how they lived, what kind of elements they form and how their lives ended.

This is the premise of what we do in galactic archaeology. We try to find the second generation of stars, which exist until the present day. And we find them in the Milky Way because they were much lower mass than the first stars, meaning that they have lived much longer than the first stars. And so when we find them and study their chemical compositions, we can infer directly not only what was the environment of the first stars, but also what are the different chemical enrichment events that happened in the early Universe.

The second generation of stars had the very first heavy elements that any star could have, and they kept forming more and more elements at their cores. When they died through the supernova explosions, they ejected these elements into the next generation. Thus during one generation after another, stars keep on enriching the Universe with more and more elements until the present day, until we get to recent stars like our sun, which has so much of these heavy elements. This is happening because of the cosmic recycling of elements.

Below: Messier 5, photographed by Hubble, is an ancient globular cluster in the Milky Way. The majority of its stars formed more than 12 billion years ago, but there are some unexpected (blue) newcomers on the scene, adding some vitality to this ageing population.

Below: Messier 5, photographed by Hubble, is an ancient globular cluster in the Milky Way.

CLASSIFYING STARS

Astronomers classify stars based on their spectral characteristic – electromagnetic radiation from the star is analysed to determine the ionisation state, giving an objective measure of the photosphere's temperature and indicating the star's colour at the peak of its spectrum. Most stars are classified as 'main sequence' using the letters O, B, A, F, G, K and M, a sequence from the hottest (O type) to the coolest (M type).

The Hertzsprung–Russell diagram (abbreviated as HR diagram) is a scatter plot of stars showing the relationship between the stars' absolute magnitudes or luminosities versus their stellar classifications or effective temperatures. The diagram was created independently around 1910 by Ejnar Hertzsprung and Henry Norris Russell, and represented a major step towards an understanding of stellar evolution.

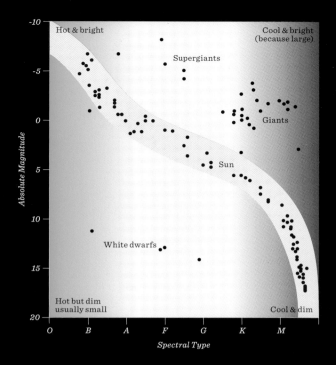

MAIN SEQUENCE STARS							GIANT STARS	WHITE DWARFS	SUPERGIANT STARS
The majority of stars in the Universe, including the Sun							*Low mass stars near the end of life*	*Dying remnant of an imploded star*	*High mass stars near the end of life*
Spectral type	O	B	A	F	G	K	M	Mainly G, K or M	D	O, B, A, F, G, K or M
Temperature	40,000 K	20,000 K	8,500 K	6,500 K	5,700 K	4,500 K	3,200 K	3,000–10,000 K	under 80,000 K	4,000–40,000 K
Radius (Sun = 1)	10	5	1.7	1.3	1	0.8	0.3	10–50	under 0.01	30–500
Mass (Sun = 1)	50	10	2	1.5	1	0.7	0.2	1–5	under 1.4	10–70
Luminosity (Sun = 1)	100,000	1,000	20	4	1	0.2	0.01	50–100	under 0.01	30,000–1,000,000
Lifetime (million years)	10	100	1,000	3,000	10,000	50,000	200,000	1,000	–	10
Abundance	0.00001%	0.1%	0.7%	2%	3.5%	8%	80%	0.4%	5%	0.0001%

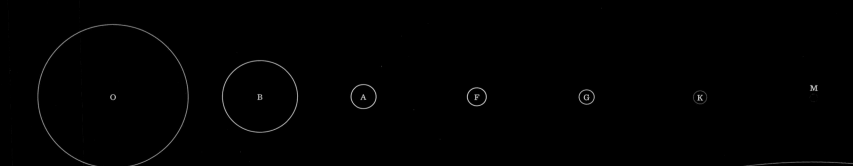

GIANT STARS

SUPERGIANT STARS

THE DARK AGES

'We found galaxies and supermassive black holes formed along the filaments fuelled by the cosmic web. This is the first time we really see the filaments connecting a number of galaxies.'
Hideki Umehata, RIKEN Institute

Let's go back 13.8 billion years, to a time before starlight, when the infant Universe was lost in darkness – a period known as the Cosmic Dark Ages. In this featureless sea of swirling particles, just 100 or so million years after the Big Bang, the Universe began taking its first few tentative steps from nothing to everything. This epoch is the bedrock for our universe today; the moment when dark matter was laying the groundwork for the construction of what is known as the cosmic web. Transforming from a universe filled with just two primordial ingredients – hydrogen and helium – to the one we see today, filled with over 90 individual elements all shaped into an endless variety of galaxies, stars, planets and moons.

Awash with the simplest of atoms, adrift in this empty universe, there was no reason for any of this to change, and this could have remained its story – an eternal, simple cosmos, with its form and function frozen in time just moments after its birth. So what happened? Why did it change? And where did the trigger come from to create perhaps the most profound shift in the history of our universe?

To understand this pivotal moment, we need to look beyond the simple atoms that existed at the time. They alone could not cross the line from simplicity to complexity, from darkness to light; we need to look further into the darkness to understand how these atoms became light, we need to see a structure hidden in the shadows, a mysterious structure that extends across the whole Universe, and that we are only just beginning to understand.

In October 2019, an international group of scientists led by Hideki Umehata, from the RIKEN Institute in Japan, reported that they had used the Very Large Telescope (VLT) in Chile to gain the first glimpse of an ancient structure that we have long predicted to exist but have never been able to directly observe. Peering 12 billion light years across the Universe towards a cluster of galaxies known as SSA22, the team were able to observe this ancient patch of the cosmos in such detail to see some of the oldest galaxies in the Universe, but they were also able to observe what lay between these bright islands of stars. SSA22 is a structure containing not just galaxies but also enormous quantities of gas sitting in bubble-like structures of such vastness that they are bigger than the galaxies themselves.

Looking at this patch of space through the power of this telescope has allowed us a glimpse of a grand galactic nursery, a concentration of cosmic material that has provided the perfect fertile ground for billions of stars to form. But this was not all they discovered. Using a relatively new instrument called the Multi Unit Spectroscopic Explorer (MUSE) on the VLT, they were able to glimpse three vast curved arms or filaments that they believe formed no more than 2 billion years after the Big Bang. And the galaxies and gas bubbles are lined, rather than randomly oriented along these arms. The intensity of this region is unlike anything we have ever seen, a dense conglomerate of galaxies and gas bubbles entwined together in a vast, delicate, tangled web. This, we believe, is our closest look at the structure that gave birth to the very first stars and galaxies, a structure that we have predicted but never seen before – the cosmic web.

Opposite: Map showing gas filaments (blue), detected by MUSE (left) and an image of a massive galaxy cluster from the C-EAGLE simulation (right), at the convergence of these filaments where a massive group of galaxies is assembling.

Right: The vast majority of the matter that makes up the Universe is little-understood 'dark matter', and the even less understood 'dark energy'.

69% Dark energy — 39% Total matter — 80% Dark matter — 20% Regular matter → Stars / Galaxies / Dust & gas

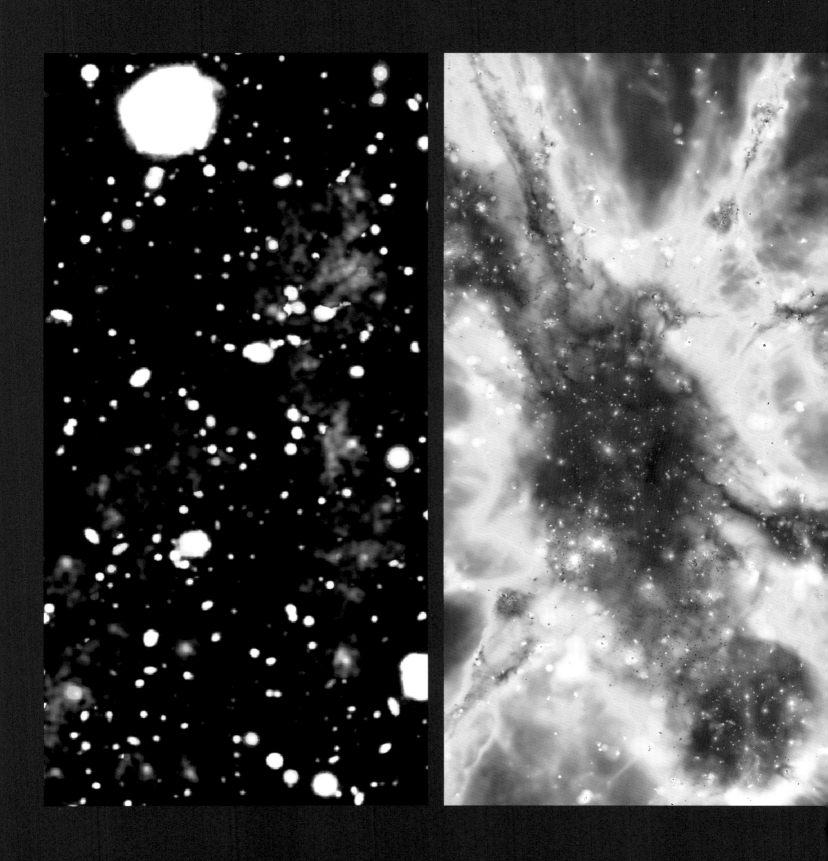

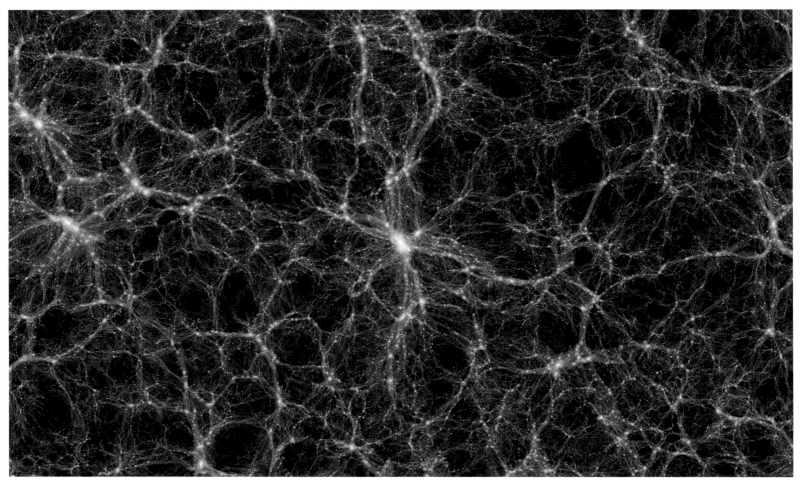

<u>Dark matter and the cosmic web</u>
The cosmic web is a term used to describe the network of filaments and halos of dark matter that have essentially been formed and collapsed over several billions of years due to the effect of gravity, as dark matter has collected in high concentrations in certain parts of the Universe and emptied out other parts of the Universe as a result. This actually forms a distribution that kind of looks like a spider's web. Hence the name.

Ultimately, it is dark matter that we think is actually allowing the process of the laying of the foundations of the cosmos to take place. But really, we don't have a full picture for what dark matter is. And depending on where you think dark matter is, the formation of the very first structures in the cosmic web will actually change. So some models of dark matter will actually stop producing structure very early on, since essentially the existence of the first dark matter particles, whereas other models of dark matter will start this at a much later point in time. And of course, this means that there will be a delay – or not, as the case may be – in the formation of the very first stars and galaxies. These are the kinds of questions that we as astronomers are asking ourselves: how can we actually try to measure and detect the very first stars and galaxies in the Universe to help us understand when the first dark matter structures were created and therefore what the nature of the dark matter might be?
Sownak Bose, Researcher, Center for Astrophysics, Harvard

The observations made in SSA22 have revealed a web that extends for over 3 million light years, made of vast, dense filaments of gas, the hydrogen and helium of the early Universe that we know provided the fuel to create the first stars. It's at the places where these filaments cross, the regions of greatest density, that we see stars, black holes and galaxies form, which suggests that this is the key to how the very first stars emerged out of the darkness of the early Universe.

Not much more than 100 million years after its birth, the Universe was still lost in the dark ages. Filled with clouds of hydrogen and helium, there were no stars to shine through the darkness, no light to reveal the heavens. But this was a universe that was changing, cooling down from the extreme temperatures of its birth that had made the existence of matter impossible in the early years. Now it was reaching a place where atoms were not only present but could actually move slowly enough to start clumping together under the growing influence of gravity.

As the Universe cooled it wasn't just ordinary matter, the hydrogen and helium, that slowly began to form structures; something else had been developing in the dark to create a hidden structure upon which the Universe could be built. Dark matter, the mysterious substance that we know exists but that remains one of the great mysteries in modern physics, had been slowly creating a network of dense structures, which were spreading across the Universe. As the dark matter cooled, the hydrogen and helium became increasingly attracted to these structures, falling into the densest regions under the irresistible pull of gravity. Gradually this gave rise to a network of strands and filaments, ordinary and dark matter entwined together and stretching across the Universe – this was the cosmic web, a web that we can still glimpse today, which created the framework to turn darkness into light. It is our universe as we know it, and it is evolving all the time, slowly reconfiguring, forming new galaxies and new stars. It was in the densest concentrations of this first forming web that the conditions were just right for the first stars to be born, to enable the Universe to take shape in a slow-motion dance.

Opposite above: This computer simulation shows the distribution of elusive dark matter in the local universe, which cannot be detected by telescopes.

Opposite below: Like a spider's web, the cosmic web is an intricate network of filaments and halos of dark matter that grows under the force of gravity.

Right: The MUSE instrument is a 3D spectrograph with a wide field of view for discovering new views of our universe, complete with a cryogenic cooling system.

A STAR
IS BORN

'A star is an energy machine.'
Kelly Korreck, Program Scientist,
Heliophysics Division, NASA

The birth of every star that has ever existed, from the Sun in our sky to the very first stars in the Universe, is based on the same basic principles of physics. Even though we have never managed to catch a glimpse of one of those earliest stars we can use these principles to help us describe them in great detail.

Around 100 million years after the Big Bang, slowly but surely, growing in the darkness at the densest intersections of the cosmic web, vast amounts of gas were being drawn together by gravity to form huge clouds, millions of light years across. Forced together under the influence of gravity and driven by the dark matter lurking in the shadows, these clouds of primordial hydrogen and helium condensed, becoming denser and hotter in certain places, until under the immense pressure the atoms themselves could no longer resist.

At some point in this early history of the Universe, two atoms of hydrogen came together in one of these enormous clouds, squeezed to a point where they could no longer exist in isolation. In this moment, these two atoms became one, fusing together to make a new element, helium, and in the process of which released an unimaginable amount of energy. Perhaps this event happened millions or even billions of times without catching the light, but in one of these clouds, one of these moments, was the beginning of a runaway process where vast amounts of hydrogen began to fuse together, releasing more and more energy until out of this maelstrom something extraordinary emerged.

As these vast clouds of gas collapsed and forced huge amounts of hydrogen to fuse, the centre of the clouds became increasingly hot, reaching temperatures of anywhere up to 10,000 degrees Kelvin; creating a protostar. A dense cloud of hydrogen atoms is sustained, with a temperature and pressure so great that nuclear fusion doesn't just sporadically occur. Almost 14 billion years ago, inside that ancient cloud of gas, the runaway fusion reaction triggered the beginning of a new structure within the Universe. As the cloud collapsed in on itself under the force of gravity, the fusion reaction created an oppositional force, an outward pressure that could act against the gravitational collapse. At first this would have been turbulent and chaotic, but as time passed, the two forces would have found an equilibrium, a balance between collapse and expansion, and that balance would have slowly allowed the cloud to transform into a stable sphere, a ball of gas, a star. As the gas and dust dispersed from around this newly formed structure, the first photons, released deep within this new structure, emerged into the Universe.

In this moment the Cosmic Dark Ages came to an end, the first star was born and the age of starlight had begun. The Universe was now filled with its first starlight; a light whose origins we are trying to understand over 13 billion years later.

We have never observed one of these first stars, but as is often the case with cosmology you don't have to see something directly to find evidence of its existence. In 2018, after years of painstaking work, a project called EDGES (Experiment to Detect the Global EoR Signature) finally provided the first hint of evidence linking us back to the moment the earliest stars in the Universe formed. Using an inconsequential-looking radio antenna no bigger than a small table, at the Murchison Radioastronomy Observatory in the red, dusty outback of Western Australia, the EDGES team was able to detect a faint radio signal that seemed to have its origin at the very moment the Universe became awash with light.

Opposite bottom: The hydrogen bomb uses the energy produced by nuclear fusion, on a much smaller scale than the fusion going on inside the Sun.

Opposite top: VIP guests watching the detonation of the early hydrogen bombs, tested in the Marshall Islands in 1952.

The impact of the first stars

The very first stars were probably much, much, much more massive than our sun is – possibly able to fit hundreds, if not up to a thousand, suns within them. They would have shone brightly in the sky and been more luminous than our sun, and because of their huge size, they likely would have appeared to be quite blue. Their high velocity would have also meant that their temperature was high, and this, coupled with their size, would have resulted in them being placed far apart from each other, sparsely dotted across the sky. Yet this first star would have been incredibly short-lived, because it was burning its hydrogen fuel so rapidly, probably existing for just a few million years before it exploded as a supernova.

Hunting for this signal required the team – led by Judd Bowman of Arizona State University – to conduct a painstaking search within the fingerprint of the Universe's origins, the Cosmic Microwave Background radiation (CMB) that we will discuss in Chapter 5. This ancient glow, formed just 380,000 years after the Big Bang, marks the moment the Universe became cool enough to transform from its hot, dense beginnings when it was filled with a plasma of electrons and protons to a place where atoms of hydrogen could exist. At that moment, the fog of the early Universe cleared and photons – light – could travel through space freely for the first time, creating the faint signal that we can still see today.

Not only is the CMB the earliest form of light we can see in the history of the Universe, a light that existed before there were even stars, it is also a benchmark against which we can search for ancient events. The EDGES experiment has been looking for the tiniest variations in the CMB that might be the remnants of these events lurking in the deep past. The theory is that as the first massive blue stars lit up, the sudden release of energy would have sent a tremor through the vast swathes of newly formed hydrogen atoms that filled the early Universe. As these stars lit up the surrounding gas, the hydrogen atoms were excited, causing them to start absorbing radiation from the Cosmic Microwave Background at a characteristic wavelength.

If this is correct, the physics suggests that the moment of first light, the cosmic dawn, should have left a trace in the CMB as a specific dip at a specific point in the spectrum of light of which the CMB is made up, a shadow caught in the CMB by the light of those stars. The theory is one thing, proving it is much more difficult.

Below: This tiny antenna on a large sheet in the remote outback has allowed the EDGES team to discover ancient signals in the Universe.

Opposite: This illustration shows what star formation may have looked like in the early Universe, as bright, blue-white stars formed in primordial galaxies.

THE PROTON-PROTON CHAIN

When two protons (^{1}H) meet in the Sun, they react, and split into two protons. On rare occasions, a deuteron (^{2}H) is produced, and a positron and a neutrino are emitted. These reactions account for 10.4% of the Sun's energy.

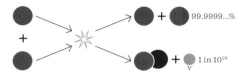

HYDROGEN-HELIUM FUSION IN THE SUN

The deuteron (^{2}H) created in this proton-proton chain easily combines with another proton to fuse into Helium-3 (^{3}He), accounting for 39.5% of the Sun's total energy. The 2 ^{3}Hes fuse with each other to create Helium-4 (^{4}He), which accounts for 39.3% of the Sun's total energy.

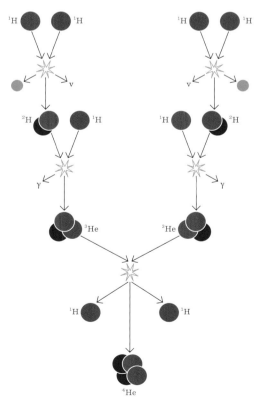

- ● Protons
- ● Neutrons
- ● Positrons
- γ Gamma ray
- ν Neutrino

The Universe is awash with radiowaves created by endless different sources in our own galaxy; the signal the EDGES team were searching for has also been stretched over the eons of time, redshifted away from the original predictable wavelength by the expansion of the Universe. Filtering out all the noise and finding the exact part of the spectrum to search for the redshifted signal required the team to filter out 99.99 per cent of the background radiation and focus in on a signal that is just 0.01 per cent of the contaminating radio noise coming from our own galaxy. Such a technical challenge would suggest the most sophisticated of technology, but looks can sometimes be deceiving in science, and although the EDGES antenna appears about as impressive as a cheap garden table, its position in the remote Western Australian outback, protected from the radiowave pollution of human activity, enabled Bowman and the team to hone in on that ancient signal from a long-gone universe and 'see' within the background radiation the moment the Universe went from light to dark. A moment that we now think occurred 13.6 billion years ago just 180 million years after the Big Bang.

It would be another 9 billion years before the most familiar of stars, our sun, would make its appearance in a light-filled, crowded universe. Although it was just one star, in one galaxy, among billions of stars within billions of galaxies, be in no doubt that the story of our star – of all stars – began at this moment. The birth of those first giant blue stars, at least 100 times bigger than our own sun, the biggest stars that have ever existed, marked the origins of every star in the Universe; the first generation of stellar gods ruling over the primitive Universe with a violent tempestuousness, a volatility that would set the course of all our history.

CARBON CREATION

Once ^{4}He has formed inside a star, two molecules can fuse to create beryllium, which can then fuse with another ^{4}He to form carbon, the building block of all life on Earth.

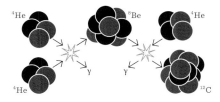

^{4}He ^{8}Be ^{4}He

^{4}He   γ   γ ^{12}C

● Protons
● Neutrons
γ Gamma ray

Deep in the heart of those first stars a process was taking place that would change the character of the Universe forever. As hydrogen atoms fused together deep inside these stellar cores, a cascade of events began that would ultimately create the building blocks of the Universe, the matter that makes up every planet and moon and everything upon them, including every part of every living thing, the very matter that you are made of.

Hydrogen is the simplest and lightest element, with just a single proton in its nucleus, but smash together four hydrogen atoms under the colossal temperatures and pressures deep in the core of a star and the result is not only the release of huge amounts of energy but also the creation of a new heavier atom, helium, with two protons in its nucleus. In those very first stars the process of fusing hydrogen into helium was driven with an intensity beyond that found in any star we can see today. Models suggest these first stars were up to 1,000 times more massive than the Sun – vast in volume and millions of times brighter than our own star. With a surface temperature of up to 110,000 degrees Kelvin, 20 times the surface temperature of the Sun (at 5,780 degrees Kelvin), the rate of hydrogen fusion to produce these levels of energy would have been immense.

But it wasn't just hydrogen that was producing these vast amounts of energy; as the levels of helium increased in these stars, the process of fusion would have taken its next step. We don't know for certain when in the life cycle of the star helium fusion would have begun, but we do know that at some point when the hydrogen supplies ran low, helium fusion would have taken over as the main source of energy fuelling the star. When helium fuses in a set of reactions known as the triple-alpha process, three helium-4 nuclei can be transformed via a cascade of reactions into a new heavier element, a carbon nucleus, which until this process took place in the earliest stars would not have existed in the Universe. At some point in the life cycle of those first massive blue stars, three helium nuclei would have come together to create an atom with six protons that would go on to change the history of the Universe. Carbon, the foundation for all life on our planet and, as far as we know, all life that could exist in the Universe, came into existence in this way, carbon that around 9.5 billion years later would spark into life and a few years after that spark into consciousness. Whether indirectly or directly, we are the product of those first stars made conscious, beings of carbon that can only exist because of our stellar ancestors.

But this was not the end of the alchemy that went on in those first stars. As the stars grew hotter, the fusion of atoms passed from one element to another, creating ever more heavy elements – nitrogen, oxygen, sodium, potassium, and one by one the atoms of the periodic table were born in the heart of those first stars, the ingredients that would eventually breathe life into the Universe, into our planet, into us.

The Big Bang wasn't the birth of the Universe, it was its conception. Only after millions of years, slowly developing in the dark, would we see the start of the Universe as we know it. The birth of the first stars marked the moment the Universe became enriched with the elements beyond hydrogen, but as those first stars burned all of those elements, all the potential they held remained locked away in the heart of those first stars, and there they would have remained were it not for another of their characteristics. Because those stars, burning bright, hot and furious, lived fast and died young, and with that mortality came the greatest of endings.

SEEDING THE UNIVERSE

At a time when most stars are just beginning their life stories, the first stars that lit up the Universe in a blaze of blue were nearing their demise. Just 2 or 3 million years after their birth, these stars were out of control. The balance between inward and outward forces could no longer hold and stability was replaced by instability. In the heart of these stars the fuel that had driven the fusion processes was depleting, no longer able to sustain the production of ever-heavier elements. Instead, another dynamic would take over. With dwindling fusion and the outward pressure that comes with it, the hydrostatic equilibrium, the balance, the inward and outward pressure that makes a star a sphere, could be sustained no longer and instead the force of gravity took over. The core, now filled with its bounty of new elements, began to collapse under its own weight, crushing the contents in on itself at a speed of collapse that is almost unimaginable. With velocities reaching up to 70,000 kilometres per second, the core of these first stars collapsed in less than a quarter of a second, resulting in an equally rapid increase in temperature and density.

STELLAR EVOLUTION

Stars are formed in nebulae. Nuclear reactions in their cores make them shine brightly, as 'main-sequence' stars, like the Sun. Eventually, the hydrogen fuel that powers these reactions begins to run out, and the stars expand, cool and change colour to become red giants. Next, small stars like our sun will shed their outer layers into planetary nebulae and become white dwarfs, while massive stars explode as supernovae, leaving behind either very dense neutron stars, or black holes.

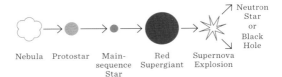

Nebula   Protostar   Main-   Red   Supernova
                     sequence   Supergiant   Explosion
                     Star

Neutron
Star
or
Black
Hole

In our universe this cataclysmic moment can trigger a series of different outcomes, including the creation of black holes and neutron stars. Exactly what happened to those first stars when their cores collapsed with such violence is uncertain, and without knowing the exact size and makeup of these stars it is difficult to apply the physics of today to that moment. But what we can suggest with some certainty is that in some of those first stars this process of collapse resulted in the core being squashed so tight it would have been no bigger in diameter than 30 kilometres, with a density akin to the inside of an atomic nucleus. If the Earth's density was increased to the average nuclear density, for comparison, its radius would be only about 200 metres, 30,000 times smaller than it actually is. However, gravity can only push so far, and the progress of this massive inward collapse would have eventually come to a halt. With temperatures reaching at least 100 billion degrees Kelvin, the conditions within this tiny stellar core are, to say the least, intense. The fusion process that had slowly halted as the star ran out of fuel is suddenly and violently reignited by these new extreme conditions, and in the briefest of moments the core rebounds, matter is almost instantaneously transformed into energy, and the Universe's greatest show, a supernova, would have illuminated the darkness for the very first time.

In death the first stars began to remake the Universe, blasting out the remnants of their lives in supernovae, transforming a binary universe of hydrogen and helium into a cosmos rich with a multitude of ingredients. These ingredients would within a few million years go on to form new stars, stars that were formed not just from hydrogen and helium but a growing list of other trace elements – known as metals – to create a new second generation of stars (Population II stars) to fill our universe.

Today, from our vantage point here on Earth, we can raise our telescopes to the heavens and see some of these oldest of stars, stars that have shone for almost as long as the Universe has existed, stars like the snappily named HE 1523-0901, a red giant star 7,500 light years from Earth in the far reaches of our galaxy that is thought to be around 13.2 billion years old (plus or minus 2 billion years!). Relatively poor in metal constituents compared to a star like our own sun, we think that HE 1523-0901 is so old it may well have been formed directly from the remnants blasted out by that first generation of blue stars that lit up the Universe. This is a star that has witnessed almost the whole history of the Universe and will go on burning long into the future until it finally cools, loses its outer layers and becomes a white dwarf, the remains of this star's stellar core, that will continue to dimly shine out in the Universe until the end of time.

Since those first stars that emerged 180 million years after the Big Bang, many generations of stars have come and gone, living their lives, enriching the early galaxies with new elements to build ever more varied stars. Stars of different sizes, temperatures and colours, all different in some way but all part of one family. The story of this grand family is one of birth and death, creation and destruction, a tale that started 13 billion years ago but that has only barely begun.

Among the billions of stars, one of them – a yellow star, on the small side – in the outer suburbs of a regular spiral galaxy has organised the rich chemistry of its nursery and put it to spectacular use: nurturing the only known life in the Universe.

Opposite: The death in 1054 CE of a star in the constellation Taurus has left behind a superdense neutron star spewing out high-energy particles.

Above: Digitised Sky Survey image of the oldest star in our galaxy – HD 140283, 190 light years away and about 14.5 billion years old.

BIRTH OF OUR SUN

Look up on a sunny day and the star you see filling our world with light and warmth, our sun, is a direct descendant of those first blue stars, the grandchild of violent ancestors that lived and died in those initial moments of starlight. When they died the newly formed atoms that forged in the heart of those stars were blasted across the Universe, beginning a journey that would lead some of those atoms into the hearts of new stars, where they would once again become part of the grandest cycle of life – a regeneration that has taken place countless times in the history of our universe.

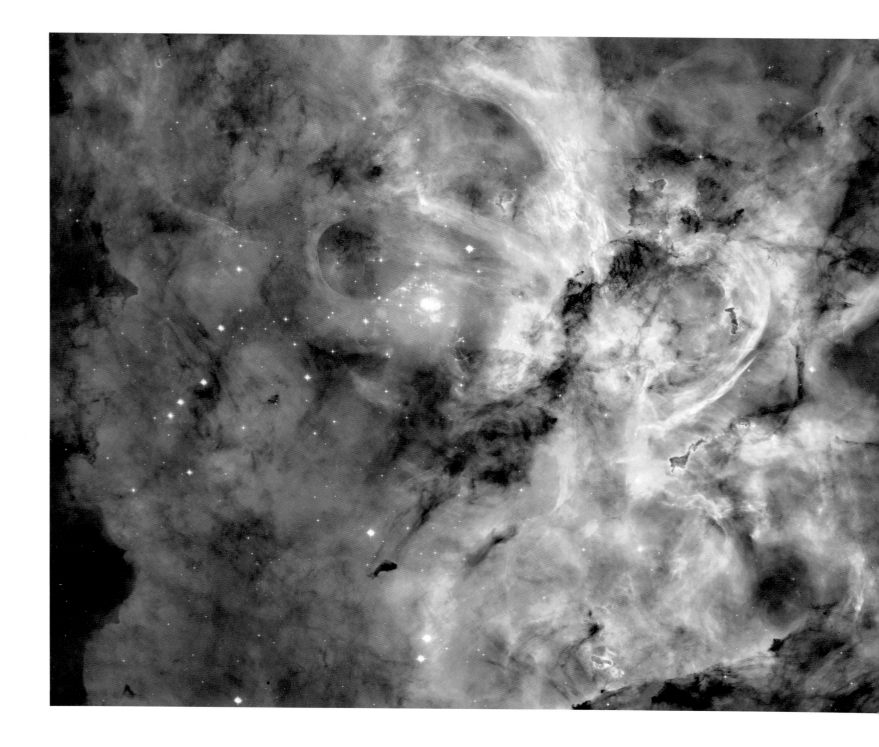

'The adage live fast, die young applies to the most massive, luminous stars, so it's the smaller, more quiescent stars that become the greatest historians of the Universe.'
Grant Tremblay, Astrophysicist, Harvard Smithsonian Center

Lost in this vast cycle of birth, life and death are the direct ancestors of our sun. We can glimpse where it all began in the lives of those first blue stars, but we will never know how many other stars have lived and died to create the collection of atoms that ended up swirling together in the outer reaches of our galaxy 5 billion years ago. All we know is that around 4.5 billion years ago this vast cloud of dust, perhaps nudged by the shockwave of a distant supernova explosion, began to collapse under its own weight, and out of the ashes of thousands of dead stars our sun was born.

Just like each of its ancient ancestors, the Sun formed through a process of gravitational collapse, slowly emerging out of a giant molecular cloud composed almost entirely of the hydrogen and helium formed in the Big Bang. Only a tiny fraction of the cloud (perhaps around 1 per cent) was made up of the heavier elements, the atoms forged through the generations of stars that lived and died before our sun was even a twinkle in the sky, but it was that 1 per cent that helped create the precise characteristics of this newly forming star.

At first there was nothing more than a spinning disc of gas and dust, contracting into a dense, hot core within which a protostar could form. An embryonic star hungrily gathering more mass from the surrounding cloud, feeding on it for half a billion years before it could feed no more. This was still a dark star, no photons could escape to fill the darkness with light, hydrogen fusion had yet to begin and, deep in its interior, temperatures had yet to climb. But eventually this would be a protostar no more, and in its moment of birth, the shroud of dust and gas was blown away and the first photons of light from our sun made their way into the dark Universe beyond.

This story of the birth of our sun is one that we could have never witnessed but that we have seen reflected in the lives of other stars that we have been able to glimpse in the darkness. Stars like HBC 1, a young star captured in the moments after its birth by the Hubble Space Telescope, the wisps of the cloud from which it was formed still clinging around it. An image of HBC 1 was captured at a stage known as the pre-main sequence, an adolescent moment between the embryonic protostar and the adult life of a main-sequence star. Our sun was once like this, held in a moment of transition on its journey to maturity. Only when the vast mass of these young stars contracts even further can the temperature rise high enough to trigger hydrogen fusion, and with that the life of the star has truly begun.

This is how our sun was born, an unremarkable main-sequence star in a seemingly unremarkable part of an unremarkable galaxy. Like the thousands of other main-sequence stars we have observed in the night sky, the life story of the Sun was determined at the moment of its birth – its mass the ultimate predictor of the life it would lead. Too big and a star will burn out quickly; a star born ten times the mass of our sun can burn itself out in around 20 million years. At the other end of the scale we find the longest-lived stars, the runts of the litter, red dwarfs that are just half the mass of our sun that will, we think, last 100 billion years before they fade into the darkness.

The lifetime of a star is not only defined by its mass but its brightness and colour as well. The bigger the star, the brighter and bluer it shines; the smaller it is, the dimmer and redder it glows. Somewhere in the middle of all of this we find our sun, a yellow star on the smallish side that from the moment of its birth was destined to live out its life over 10 billion years.

The Sun began its life alone, surrounded by the remnants of the cloud of gas and dust from which it had formed, and for the first million years of its life the empire of the Sun lacked any worlds to shine on. It would take millions of years for the tiniest of particles to be drawn together, for dust to become rocks, rocks to become boulders, and finally through an endless game of collision and accumulation, the planets began to emerge from the clouds. These new worlds were built from atoms that had been created inside the ancient ancestors of our sun, long-dead star dust reshaped into a system of planets around a newly born star.

Previous page: In the Carina spiral arm of the Milky Way lies NGC 3603, a busy region of stars that exploded into existence 2 million years ago.

Above: This Chandra image of the region NGC 3603 in the Carina spiral arm of the Milky Way Galaxy reveals dozens of extremely massive stars born in a burst of star formation about 2 million years ago.

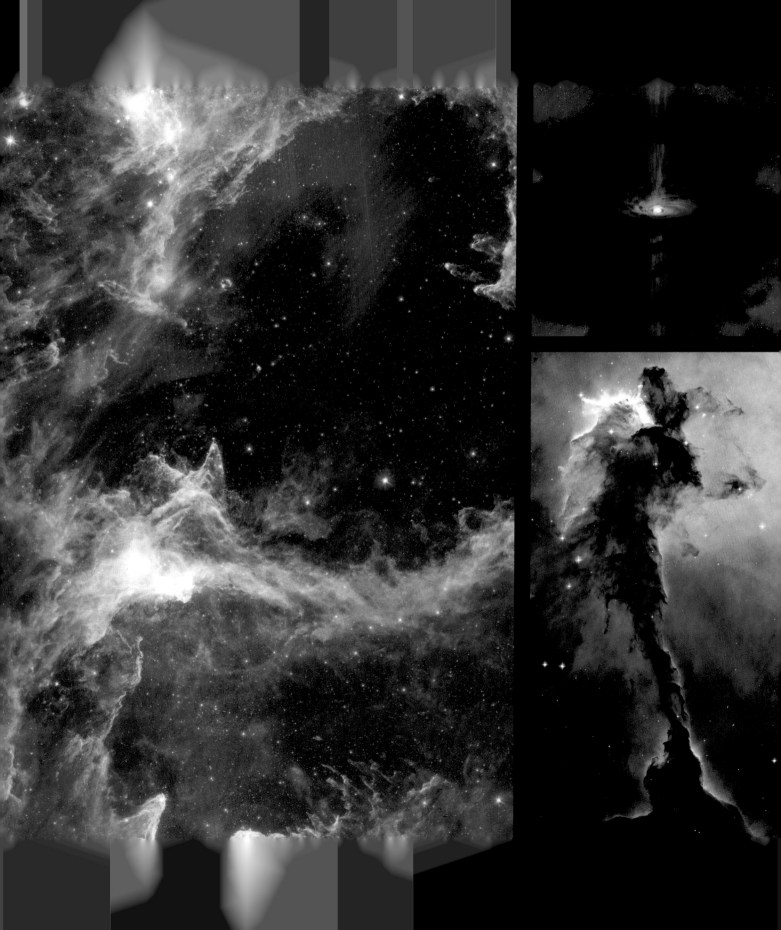

This craggy fantasy mountaintop enshrouded by wispy clouds looks like a bizarre landscape from Tolkien's *The Lord of the Rings* or a Dr Seuss book, depending on your imagination. This NASA Hubble Space Telescope image, which is even more dramatic than fiction, captures the chaotic activity atop a three-light-year-tall pillar of gas and dust that is being eaten away by the brilliant light from nearby bright stars. The pillar is also being assaulted from within, as infant stars buried inside it fire off jets of gas that can be seen streaming from towering peaks.

This turbulent cosmic pinnacle lies within a tempestuous stellar nursery called the Carina nebula, located 7,500 light years away in the southern constellation Carina. The image celebrates the 20th anniversary of Hubble's launch and deployment into an orbit around Earth.

Scorching radiation and fast winds (streams of charged particles) from super-hot newborn stars in the nebula are shaping and compressing the pillar, causing new stars to form within it.

Streamers of hot ionised gas can be seen flowing off the ridges of the structure, and wispy veils of gas and dust, illuminated by starlight, float around its towering peaks. The denser parts of the pillar are resisting being eroded by radiation much like a towering butte in Utah's Monument Valley withstands erosion by water and wind.

Nestled inside this dense mountain are fledgling stars. Long streamers of gas can be seen shooting in opposite directions off the pedestal at the top of the image. Another pair of jets is visible at another peak near the centre of the image. These jets (known as HH 901 and HH 902, respectively) are the signpost for new star birth. The jets are launched by swirling discs around the young stars, which allow material to slowly accrete onto the stars' surfaces. Hubble's Wide Field Camera 3 observed the pillar on 1 and 2 February 2010. The colours in this composite image correspond to the glow of oxygen (blue), hydrogen and nitrogen (green), and sulphur (red).

Opposite: This NASA Hubble Space Telescope image captures a giant pillar of gas and dust in the Eagle Nebula, within which infant stars fire off jets of gas.

Below: Taken by the Hubble Space Telescope, these four images reveal the birth of stars in the Orion complex, nestled in the gas and dust.

ORION

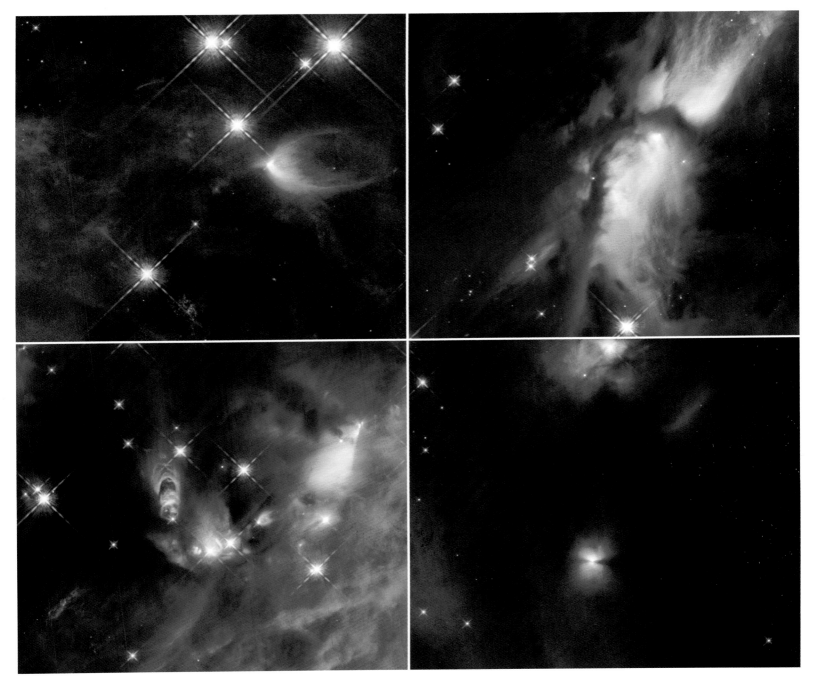

ENGINE

As our solar system emerged from the shadows 4.5 billion years ago, a dance began between our star and its planets that continues to this day. Containing 99.8 per cent of all the mass in the Solar System, the Sun has not only dominated its realm through its size but also through its power. From the very beginning the Sun's energy has shaped the worlds around it, raining down energy onto the four rocky planets – Mercury, Venus, Earth and Mars. The Sun is an absolute ruler, wielding the power of life and death over everything in its domain.

Mercury is a scorched planet, so close to the Sun that daytime temperatures can reach 430 degrees Celsius, dropping to -180 degrees Celsius at night, making it a planet unlikely to support any life that we know. Scientists believe that Venus, similar in size and structure to Earth and often called our planet's twin, was once

Left: The Earth reflects the Sun's energy as thermal long-wave radiation. The blue swathes represent thick clouds, the tops of which are so high they are among the coldest places on Earth.

THE ENERGY BUDGET

Energy from the Sun is balanced by energy reflected or emitted from the Earth. Of the total solar energy that reaches Earth, 29% is reflected straight back into space, 23% is absorbed by the atmosphere and 48% is absorbed by the Earth. The Earth emits energy into the atmosphere as thermals, evapotranspiration and heat, which the atmosphere then emits back into space, balancing the Sun's original input. Though a substantial amount of the heat emitted by the Earth is trapped by the atmosphere, warming the planet.

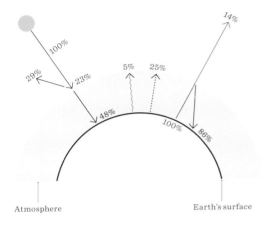

Atmosphere   Earth's surface

〜〜〜〜〜 Thermals *(conduction/convection)*

.... Evapotranspiration *(latent heat)*

a water world and a possible haven for life, but today, with its surface temperature of 465 degrees Celsius and hurricane-force winds that blow acidic clouds around the planet every four days, it is a toxic and boiling hell, the hottest planet in our solar system. The fourth planet from the Sun, Mars is a world with seasons, and one where, explorations have shown us, billions of years ago oceans flowed for millions of years, but now it is a cold, dusty, dead world.

All of this driven by the power of the Sun, the influence of a god that can strip a world of its oceans, whip up an atmosphere to destruction and heat a surface until it melts. In the dance of the planets only the Earth seems to have escaped with the gentlest of touches from this all-powerful god, as the only world to retain its oceans, to keep its atmosphere in check and its temperatures under control so that life as we know it can exist and thrive. But even here on Earth the power of our star to create and destroy can be seen written everywhere across its surface and is witnessed every single day. It's the Sun that drives the movement of water around our planet; its heat causes liquid and frozen water to evaporate, lifting tonnes of water from the oceans each day to drive the water cycle that over millions of years has carved the great canyons and valleys that shape the surface of the Earth. It's the Sun that drives our atmosphere to create the ever-changing weather that defines our world – from hurricanes to monsoons, from tornadoes to blizzards, all of this is powered by the fusion of hydrogen in a star 150 million kilometres away from Earth.

Just as the Sun is the powerhouse of our planet, it is also the driving force behind our solar system – without it the planets would be not much more than barren rocks adrift in the hard cold of space, but with it these inhospitable rocks are transformed into worlds, dynamic living worlds that are shaped by an ever-changing climate, driving the seasons and zones we see not just on Earth but on the other rocky planets as well. On Mars we see the melting of the ice caps each summer and the vast dust storms that blow up and envelop huge swathes of the planet; on Venus we have seen above its acid clouds, enormous storm systems that whip around in the higher atmosphere; and even in the farthest reaches of the Solar System, 4.5 billion kilometres from the Sun, we see active geology – ice geysers driven by the heat of the Sun on Neptune's icy moon, Triton.

All of this power to create change, to generate the dynamism of a living planet, emerges from a beautiful law of physics. In the simplest terms it says that if you have a temperature difference, you can do something called 'work'. In physics, work is a technical term that describes 'the measure of energy transfer that occurs when an object is moved over a distance by an external force at least part of which is applied in the direction of the displacement' – essentially, that energy has been transferred. For our purposes this means that you can create change, create dynamism – such as compress a gas, or cause motions within a body through an external magnetic force. It's why the fires we lit inside the machines of the Industrial Revolution were able to create such change – power emerges from the transference of energy from a hot to a cold point in any given system. In an engine, that difference in temperature is what ultimately allows a piston to be driven and a motor to do 'work'. In the Sun's Earth system the cold thing is space and the hot thing – the fire – is the Sun. Without that temperature difference no work could be done – no wind would flow, no rain would fall anywhere in the Solar System. The Sun, like all stars, is an engine of creation, our engine of creation, that we are only just getting to know properly.

Left: The Sun is the Earth's powerhouse, driving life on our planet every day.

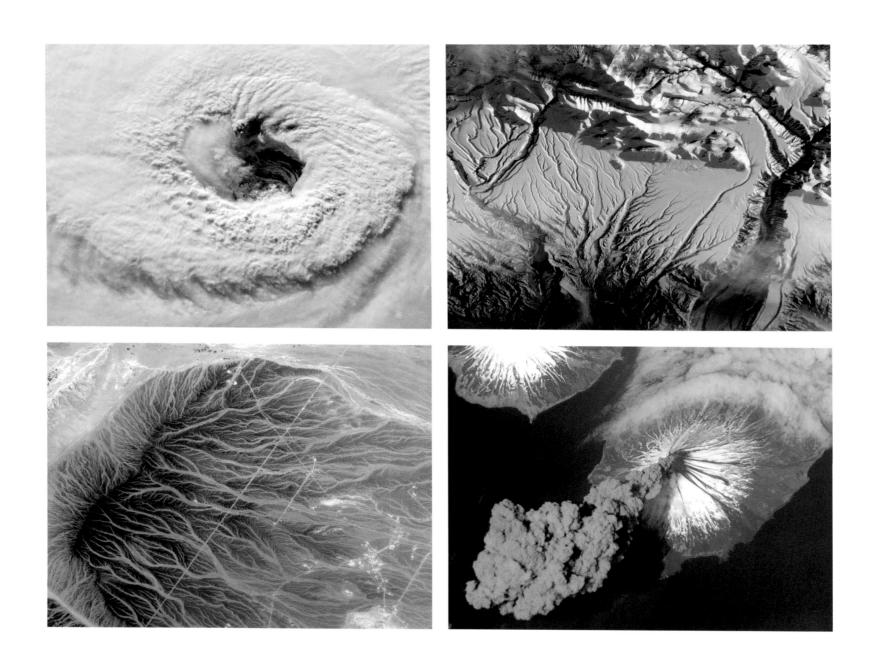

Above: Here on Earth the work of the Sun is evident everywhere on our planet – from carving canyons and valleys, driving the water cycle from the seas into the sky, and whipping up the ever-changing weather, from monsoons to hurricanes, tornados to blizzards.

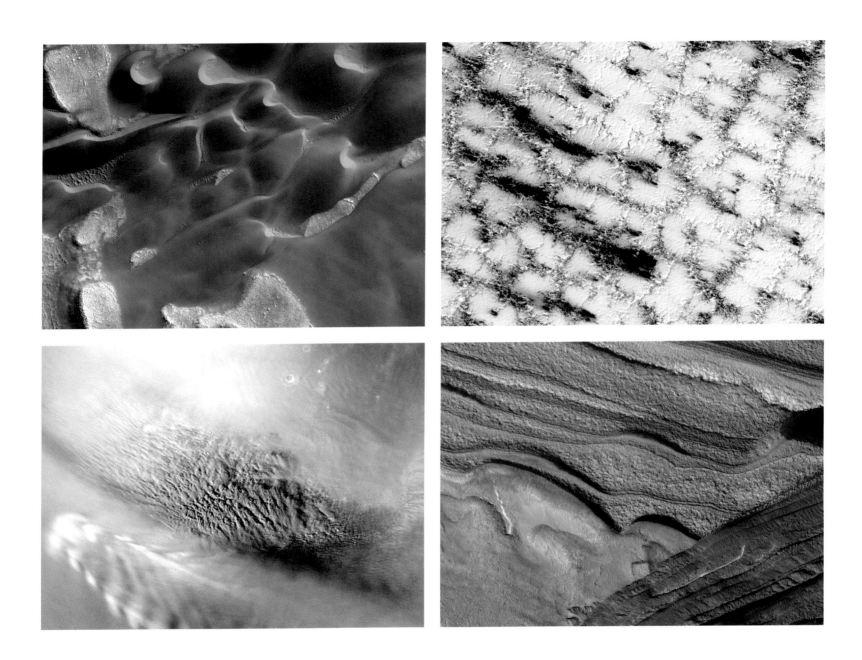

Above: Mars is long dead after the loss of its atmosphere, its surface ravaged by the Sun's rays, with melting ice caps and vast dust storms that envelop the planet.

TOUCHING
A STAR

For centuries we've looked up and tried to understand the mystery of the Sun, but only in the last 60 years have we been able to reach out and explore our star with an ever more daring armada of probes, taking us closer to the extremes of the Sun, to observe it, test it and ultimately touch it in our pursuit of understanding the full extent of its power.

The first of our solar explorers was the Pioneer 5 probe, launched on 11 March 1960. This tiny spin-stabilised probe was just 43 kilograms in mass and just over half a metre in diameter, but despite its diminutive dimensions it left the planet from Cape Canaveral Air Force Station with lofty ambitions, a set of mission objectives that would take us closer to the Sun than ever before. Originally designed with a mission plan to take it on a flyby of Venus, the initial launch date was delayed and so, with the window of opportunity missed, the flight plan was redesigned for a direct entry into solar orbit between Venus and the Sun. Over a period of 140 hours the probe sent back data from interplanetary space, sending 3 megabits of data across 36.4 million kilometres of space to be received by the Lovell Telescope at Jodrell Bank and the Kaena Point Satellite Tracking Station on the island of Oahu, Hawaii. Its precious data wasn't received until 30 April 1960 and with it came confirmation of a long-suspected secret of our sun. The Solar System is filled with a vast magnetic field, the heliosphere, which stretches for billions of kilometres, way beyond Pluto. It is created by the solar wind dragging out the Sun's magnetic field and influences all the worlds it touches. Here on Earth it interacts with our own magnetic field, creating space weather and phenomena such as the aurora borealis.

After the success of the Pioneer 5 probe, NASA sent a fleet of spacecraft to peer at the Sun from afar. Pioneers 6, 7, 8 and 9 all went into orbit around the Sun at a distance similar to the Earth (Earth being 1.0 AU away, and the probes orbiting at a distance of between 0.8 and 1.1 AU), all with the ambition to observe and measure the characteristics of our star in ever more detail. After last contact with Pioneer 9 in May 1983, observation of the Sun continued from afar through NASA's Solar Maximum Mission, Japan's Yohkoh and the European Space Agency's Ulysses, but it wasn't until 1995 with the launch of SOHO (Solar and Heliospheric Observatory) that our ability to image and explore the Sun really took a giant leap forward. A collaboration between ESA and NASA, SOHO was only expected to complete a two-year mission plan, but after more than 25 years in space this solar explorer is still revealing new information about the Sun, allowing us to explore from deep within its core all the way to the outer layer of the corona. The instruments on the spacecraft have also provided a vast array of new images of the Sun that have revealed the violence and beauty of its incandescent surface. The biggest solar flares we've ever recorded have been captured by SOHO, explosions that we've been able to not just observe but measure as the explosions on the Sun's surface hurl billions of tonnes of magnetised gas into space at speeds up to 7.2 million kilometres per hour.

Above: Venus, entirely
wrapped in unbroken clouds,
as captured by Pioneer on
its flyby en route to the Sun.

Above and right: NASA's Pioneer 5 being readied for launch in 1960, and imagined in space, heading towards the Sun.

Top: Pioneer 5 reached heliocentric orbit between Earth and Venus to create the first map of the Sun's magnetic field.

OUR SOLAR BUBBLE MOVING THROUGH SPACE
Space inside the heliosphere (the bubble around our solar system created by the Sun's magnetic fields and the solar wind) is hotter than the space around it. Red indicates temperatures of about 1 million degrees Kelvin. Black lines indicate the flow of the solar wind inside our solar bubble and the flow of the interstellar wind outside.

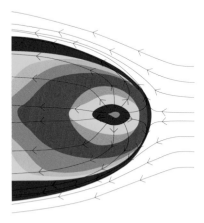

But for all the new understanding and wonder that spacecraft such as SOHO have created, it's only very recently that we've had the audacity and the technology to not only look towards the Sun but actually attempt to touch it as well. The Sun contains more than 99 per cent of the mass of the Solar System. Its surface area is nearly 12,000 times bigger than that of Earth, while its mass is 330,000 times greater.

Launched on 12 August 2018, the Parker Solar Probe will dare to take us closer to our star than ever before. At the time of writing it has already gone closer to the Sun than any other human-made object, coming within 13.5 million kilometres of the solar surface at its closest approach on 17 January 2021. This is just the beginning, though, because over its seven-year mission its trajectory will take it on an annual flyby of Venus, each time using the planet to shift its path into an ever-smaller elliptical orbit. By 2025, when it makes its twenty-fourth orbit around the Sun, it will reach an altitude of just over 6 million kilometres above the Sun's surface and fly inside the atmospheric region known as the corona for the very first time – in effect we will be touching a star, our Sun, for the very first time.

For all of our understanding of the Sun, much about it still remains a mystery. The temperature at its surface is 5,500 degrees Celsius, yet the clouds of plasma that rage above it, the corona, are over 1 million degrees Celsius.

Coming this close to a star will provide extraordinary new opportunities to understand many of the Sun's mysteries, with Parker's mission objectives aiming to allow us to explore the mechanism by which the corona becomes superheated, how

Right: The heliospheric current sheet is the surface in the Solar System where the polarity of the Sun's magnetic field changes from north to south, creating a field extending throughout the Sun's equatorial plane in the heliosphere. The shape of the current sheet is created by the Sun's rotating magnetic field interacting with the solar wind. The thickness of the current sheet is about 10,000 km near the orbit of the Earth.

'The Parker Solar Probe is revolutionary because it's the first time we're actually going 94 per cent of the way from the Earth to the Sun to experience the solar corona; it goes screaming towards the Sun at 400,000 miles per hour, making it the fastest human-made object.'
Kelly Korreck, Program Scientist, NASA

the solar wind is generated and propelled across the Solar System and how the complex solar magnetic field functions, as well as unprecedented images from being so close to a star.

However, with all this new opportunity comes risk; Parker will be exposed to conditions never before endured by any machine – inside this part of the Sun's hostile atmosphere temperatures will exceed 1 million degrees Celsius. Combined with an onslaught of high-energy particles and an endless violent magnetic storm, this is not an environment conducive to the function of delicate scientific instrumentation. Sitting between Parker and its destruction is just 115 millimetres of the Thermal Protection System, or TPS, a 2.5-metre diameter heat shield that will keep the body of the spacecraft at a balmy 30 degrees Celsius while temperatures rise to 1,400 degrees Celsius just on the other side. It's this carbon composite sandwich that will protect Parker and allow it to function in the most hostile of environments, but even with this protection and a host of other cooling design features, the orbit of the spacecraft still needs to be highly elliptical so that it only spends a brief period in the inferno before flying away from the Sun at speeds of 200 kilometres per second or 720,000 kilometres per hour, which will make it the fastest object in the history of space exploration. The mission has only just begun but already it is revealing the true character of our star. Violent and tormented, an atmosphere convulsed by eruptions and torn apart by explosions, this is no benign god, this is a god born of violence, a destroyer of worlds.

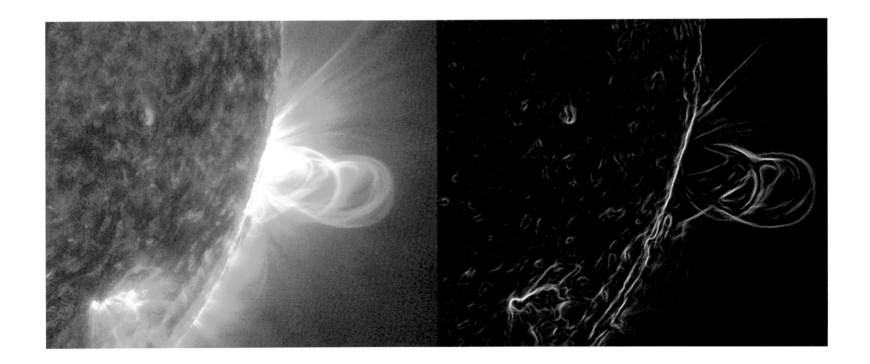

Above: Captured in July 2012, the magnetic field lines in the Sun's atmosphere twisted the coronal mass ejection into a unique slinky shape.

PARKER SOLAR PROBE

Kelly Korreck, Program Scientist, NASA

The Parker Solar Probe takes seven gravity assists from Venus in order to get into that orbit because you have to actually give up energy in order to get closer and closer to the Sun. And the only way really to do that is to have a dance with seven Venus slingshots in order to get closer. To do this, the Parker Solar Probe has to be able to travel fast – 400,000 mph – because of its orbit. The orbit is highly elliptical and using Kepler's Second Law, the probe has to speed up when it goes through that part of the orbit and then slow down as it goes further out towards Venus and Earth, then back into the Sun.

What makes Parker so great is the fact that it has a great set of instruments that work together in order to look in all directions and to experience things that we think we need to know in order to solve those big mysteries about coronal heating and the solar wind. One of the very interesting facts about this, too, is that we actually need heaters when we're closest to the Sun. So one of the engineering marvels that we had to do was to create this heat shield that blocks off heat at 25,000 degrees on the probe's approach to the Sun; but at the back end of the probe, because there's no other light coming from anywhere as you've blocked out all the Sun and it's only seeing deep space, we actually need to put a heater.

The testing of Parker Solar Probe was very extreme and we had to become very creative. There weren't many places where we could test these types of instruments. They weren't used to this level of heat and radiation. For that solar probe cup, a couple of the engineers came up with IMAX film projectors, because if you invert an IMAX film projector that's normally projected on to multiple-storey walls, then invert that into something that's focused down on something this big, you end up simulating the intensity of the Sun. So we took four of these, inverted their lenses and focused them in a chamber to test the Faraday Cup. And that was just the cup.

Everything went through different testing. The entire spacecraft went into this great big three-storey vault at Goddard Space Flight Center in Greenbelt, Maryland, to simulate deep space. There they sealed it off and sucked out all the air, then put different temperatures on it, and then they expose it to extreme space and make sure it's vacuum safe. They also put it in a room with a huge speaker and basically play the sound of launch to the spacecraft to vibrate it – an acoustic test to make sure it doesn't vibrate apart as the rocket launches.

Launch day was amazing. It didn't go off the first night because there was a valve issue. It ended up going up the second night and I knew it was going up because I was sick to my stomach. I could not sleep. The Delta IV Heavy is a very slow rocket compared to the other launches I've seen, so I just saw fireballs under something I loved and was very, very frightened for a while, then realised that this was all OK as it slowly made its way up into the sky.

Since that launch, Parker has found out some really interesting and amazing things. The Sun isn't the static ball we see from Earth, it's actually a roiling, boiling mess of gas that's constantly changing, so getting pictures and monitoring daily, if not almost by the second, has really altered our view. One discovery changes the road map that we were looking at in terms of magnetic fields. We thought that the road map would look something like a sweeping, gentle curve, but instead we're looking at these magnetic fields that basically look like switchbacks, that turn very quickly and sharply.

The other interesting thing that Parker has found is dust. We didn't think there would be dust there. It comes from the original formation of the Sun, and when the Sun formed we would have expected the dust to go away. But we still have a very dusty area that we're going through. That just doesn't have implications for our star, but for all of stellar evolution as to how that dust stays or what sticks around or doesn't stick around.

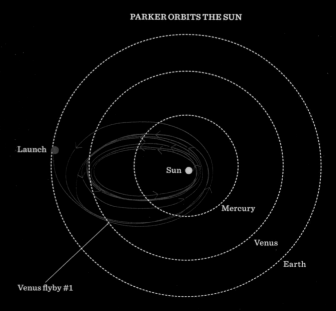

PARKER ORBITS THE SUN

Launch

Sun

Mercury

Venus

Earth

Venus flyby #1

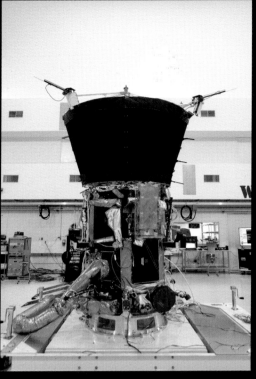

EARTH'S ENERGY SOURCE

Around 4.5 billion years ago, as the Earth emerged from the dust and gas left over from the Sun's formation, a dance began between this new planet and the star at the centre. In its infancy the Earth was a hostile and unwelcoming place, the surface was molten, heated by the endless collisions of its formation, volcanism would shape and reshape the surface as it slowly cooled, solidifying from a molten mass into a solid rocky surface over the first 800 million years of its life. We know little detail from this Hadean period of our planet's history, but we can glimpse back and imagine a world with no atmosphere, no oceans, no rocky surface and, of course, no life. Bombarded by the energy of the young Sun, this was no Eden. Without an atmosphere to act as a shield, solar radiation would have poured down onto its toxic surface, destroying any nascent life forms from ever taking hold. But at some point in those first 500 million years or so, life began on our planet, primitive life that emerged far from the toxic surface – most likely deep within the oceans. Ancestors of that simple life are still with us – the archaea, bacteria that hide even today from

'After the birth of the cosmos not much that was particularly interesting was going on, but this was the bedrock for everything that then transpired in the Universe to make it look like how it does today. It was actually being sculpted right at that time.'
Sownak Bose, Researcher, Center for Astrophysics, Harvard

destructive effects of the Sun in the ocean depths. There is no discernible reason why that simple life suddenly changed, but there must have come a day when life took just one small step that would become a giant leap in our planet's evolutionary history.

The timescales of the Universe are so vast, so beyond the imagination of the human mind, that we can easily get lost in the eons of time within which the story of the Universe is contained. The billions of years through which our planet has evolved blinds us to the fact that there are single days that change everything, moments when in an instant the Universe, the planet and all life on it can shift beyond recognition. Our planet has seen many of these days – the Earth's surface is scarred with wounds caused by meteorites that appeared on a 'Tuesday afternoon' and in an explosive instant shifted the course of all living things. But tipping points are not always so dramatic; the change may happen in an instant but the consequences can take millions, if not billions, of years to play out.

Below: Scientists believe that life on Earth may have begun in seafloor vents that spew mineral-rich water into the oceans and act as a hub for energy for water creatures.

Around 3.5 billion years ago there was one such day when the smallest of changes resulted in perhaps the greatest ever shift of life on Earth. This was a day that would have begun like any other, the Sun would have risen in the east as the Earth turned anti-clockwise on its axis. And as the Earth turned, the most primitive of life forms would have stayed hidden beneath the planet's surface, sucking its life force from Earth's internal heat around the deepest of ocean vents. This is how the planet might have stayed, how our planet could have so easily looked today – living, but essentially lifeless. And yet on that day the Universe was primed for change, over a billion years of generations of stars had enriched the galaxy with all the ingredients to make planets, and on one planet, our planet, those ingredients had combined to form living, organic matter. That life was now ready to do something extraordinary, to reach out across 150 million kilometres of space and connect with our star, creating a bond between life and light that would change everything. In this moment the Sun would transform from destroyer to creator, a moment that would breathe vast amounts of energy into life on our planet.

We can deconstruct the events of that 'day', pulling it apart into the very smallest of details. It begins 150 million kilometres from the Earth, deep within the Sun,

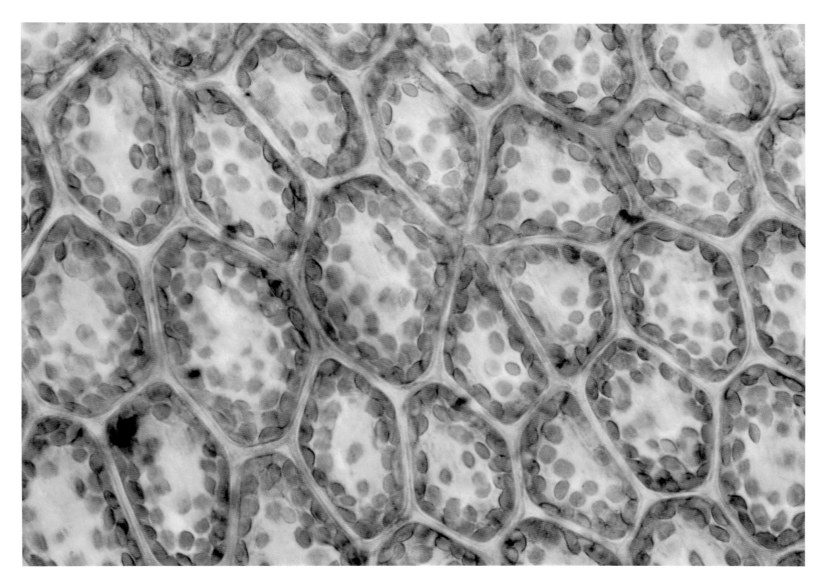

THE JOURNEY OF LIGHT

Photons are particles of pure energy, emitted as a by-product of hydrogen fusion in the Sun's core. Once created, photons must make their way to the surface of the Sun, where they escape as light. But the path to the surface is a tortured one – every time a proton collides with an atom, it is propelled in a random direction. Mathematicians call this form of motion 'a drunkard's walk' and estimate the photon will take between 5,000 and 500,000 years to reach the Sun's surface (depending on how densely populated the core is). But once it reaches the surface, it's only an 8-minute journey to the Earth.

'In some ways we take the Sun for granted. But when you think about it, the very existence of life on Earth is connected to the Sun, which makes it much harder to take it for granted.'
Nia Imara, Astrophysicist, University of California

where the fusion of hydrogen atoms results in the release of huge amounts of energy – often including the creation of a photon. Each new photon goes on its own unique journey from the Sun's interior to its surface, taking anything up to a million years from the moment of its creation (to be absorbed and emitted by countless other atoms) until it finally completes its intra-stellar voyage and leaves the surface of the Sun. Each second the number of photons released from the surface of the Sun is more than a billion billion times greater than the number of grains of sand on our planet. Streaming out from our star to illuminate every planet and moon in the Solar System, signalling our existence across the Universe.

Since the Earth first came into existence this grand stream of light has endlessly carried the energy of the Sun to the surface of our planet, illuminating and heating the surface through the daily cycle of light and dark. But around 4 billion years ago one single photon left the Sun and did something profoundly different. Journeying across the 150 million kilometres of space, travelling at almost 300,000 kilometres per second it would have taken around 8 minutes for this single photon to reach the surface of the Earth. However, unlike the billions of photons travelling with it, this photon didn't hit the land or the ocean and simply disappear. Instead, this photon shot into the ocean and encountered a new form of life, one that no longer needed the heat of the Earth to feed on.

We do not know exactly why or how, but around 3.5 billion years ago the primitive life on Earth evolved to no longer hide from the light of our star but to actually gain nourishment from it. These bacteria moved away from the ocean depths, edging nearer to the surface where light could reach them, harnessing the Sun's energy instead of hiding from it. These ancient bacteria took energy from the Sun's photons and used that energy to power a remarkable chemical transformation: taking carbon dioxide and water and turning it into a sugar, into an energy stored in chemical form – food. In this instant a new bond had been established and life was feeding directly off the energy of a star, as far as we know for the very first time in the history of the Universe. No longer plugged into the heat of the Earth's core, life had tapped into the most powerful source of energy in the Universe and on this planet that energy was everywhere. Liberating life from the depths, building the base of a vast food chain that would one day effortlessly pass the energy of the Sun from the simplest of life forms to the grandest of creatures ever to walk our planet.

The evolution of photosynthesis changed everything – not just storing the Sun's energy within the cells of simple life but also crucially producing a waste product, oxygen, that would flood into our atmosphere in vast quantities and change the energetics of almost all life on Earth. The cellular combustion of oxygen – aerobic respiration – allowed complex life to evolve from primitive algal bacteria into the complex multicellular life that we see across the planet today. Every plant, tree, fungus, animal and human is a result of the massive shift in energetics that the process of photosynthesis brought to our planet.

Trillions of stars have existed since the Universe began, but ours is the only one that we know of around which photosynthesis has evolved. The only star that, thanks to a chance mutation billions of years ago, can reach out across space every day and feed an entire planet. And that makes our sun unique in the cosmos. It is the Creator. Everyone we love. Everything we value. Our supreme accomplishments as a civilisation. All owe their existence ultimately to stars.

Opposite: Photosynthesis changed everything on our planet, bringing life and energy to our young star.

END DAY

'The Universe is like a slow-motion fireworks show. One day, somewhere in the Universe, billions and billions and billions of years from now, the last star will be born and we'll enter a new period of darkness.'
Grant Tremblay, Astrophysicist, Harvard Smithsonian Center

Today we live in an age of starlight. Our galaxy alone contains at least 100 billion stars, although the upper estimates suggest there could be as many as 400 billion suns in our local neighbourhood. Our nearest neighbouring galaxy, the Andromeda spiral, contains well over double that number. With at least 100 billion more galaxies awaiting in the Universe beyond and our conservative estimates suggesting that on average each galaxy contains 100 billion stars, that's a lot of stars. But in many ways, searching for this number, just like trying to count the endlessly compared grains of sand, is perhaps not really that useful. What we can be certain of is that the Universe is populated with a seemingly infinite number of stars and we find ourselves alive in an era of extraordinary stellar abundance, a moment in the history of our universe that will never be repeated. For thousands of years we have cast illusions of eternity into the night sky, believed the stars would live forever, but this is a moment that cannot last.

As we have watched the life stories of the stellar gods play out above our heads, we have come to realise that even gods are not immune to the ultimate truth of the Universe – everything must die. Where there is light there will be darkness, where there is growth there will be decay. Stars are creators but they are also jealous guardians of their creations; in death they do not let go of all they are made of, not everything is returned to the cosmos to be used again in the endless generations of stellar renewal. Instead they hang on to some of the elements they have made in their lifetime, hiding them away deep in their core. Dead stars become planet-sized fossils drifting through space, and as more fossils litter the Universe, more life-giving elements are locked away, slowly starving the cosmos of the material needed to make new stars.

<u>Life after Earth?</u>
Astronomers get really, really excited when we see systems like TRAPPIST, which has not one, not two, but three planets orbiting in this very special region where it could host liquid water. We get excited whenever we find any planet orbiting at this special distance. But in the case of TRAPPIST, it's especially exciting because this particular star being a small star, a low-mass star, that star is going to continue fusing hydrogen into helium in its core and continue shining light on its planets for potentially hundreds of billions of years, many, many times the age of the Universe. And so if life takes hold on any of those planets, it will have hundreds of billions of years to proliferate and to persist. If we can have planets that could potentially have water – that could potentially have life – that life could persist for hundreds of billions of years.
Phil Muirhead, Astrophysicist, Boston University

We are already at a point in the history of the Universe when more stars are dying than are being born. When we look up into the night sky, we are seeing a universe in an imperceptible decline, and despite the infinite number of stars above our heads the age of starlight is already coming to an end.

Stars won't suddenly disappear, of course, they'll still be here for trillions of years to come, but slowly, inevitably, over time our universe will grow darker, colder and emptier. Many stars will die in the eons that follow, but there are a few, a specific class of star, that will live on to bear witness to all that is to come.

Discovered in 1999, TRAPPIST-1 is a star that has shone for more than 7.5 billion years, making it nearly one-and-a-half times the age of the Sun. Located just 40 light years from Earth, this relatively nearby star has hidden in the shadows for so long due to the dimmest of light that shines from its surface. TRAPPIST-1 is a tiny star, a tenth of the size of our sun and emitting less than 0.05 per cent of the energy of the Sun, most of which is emitted in infrared, making it invisible to the naked eye from Earth. This really is the dimmest of stars, one of the best examples we have studied of what is known as a red dwarf, and an 'ultra cool' one at that. Cool and slow-burning, it has a temperature of just 2,238 degrees Celsius compared to our sun's blistering 5,500 degrees Celsius. But what makes this type of star so dull among its stellar brethren is also what makes it so remarkable. The low luminosity of red dwarf stars like TRAPPIST means that, unlike their earliest stellar ancestors, they have extraordinary longevity, burning slowly for what could be a 12-trillion-year life span, longer than any other type of star in the Universe. That's a lifespan that will extend thousands of times the age of the current Universe, a star system that will bear witness to the life story of the Universe from childhood to old age.

Even more intriguing than the slow-burning red star at the centre of this system are the seven rocky worlds that we have found orbiting it. Five of these planets are similar in size to the Earth, with the other two slightly smaller worlds sitting between Mars and our planet in scale. Unlike in our solar system, all seven of them are positioned close-in to their star, circling in a tightly packed bunch; every one of them is in orbit nearer to the star than the scorched planet Mercury is to our own sun. Not only that, but as they are close together they create a spectacle in each other's skies, multiple worlds (far bigger than the Moon appears to us) looming over their various horizons. This all means that so much of this planetary system is alien to our own; years last a handful of days, worlds are orbitally locked with one side permanently facing the cold void of space and the other side facing the star in eternal daylight.

But for all of this strangeness there is also an odd familiarity, because these worlds may live close to their star but when that star is a red dwarf, being close doesn't mean being scorched like our own battered Mercury. We have evidence to suggest that three of these planets lie within the habitable zone: TRAPPIST-1b, 1c and 1d are all in an orbit that makes them neither too hot nor too cold to allow oceans to flow across their surfaces and atmospheres to fill their skies, a tantalising possibility the Hubble Space Telescope hinted at when it turned its gaze in the direction of TRAPPIST in 2017.

All the evidence currently suggests that these planets could be a home to life, living on the surface of a planet locked in perpetual day and night, huddling for light and warmth from a faintly flickering star.

Opposite: Artist's impression of what it might be like to stand on the surface of TRAPPIST-1e and survey the planets and stars around it.

Above: This travel poster was inspired by the fact that TRAPPIST-1's planets are so close together, if you stand on one you can see its neighbouring planets above.

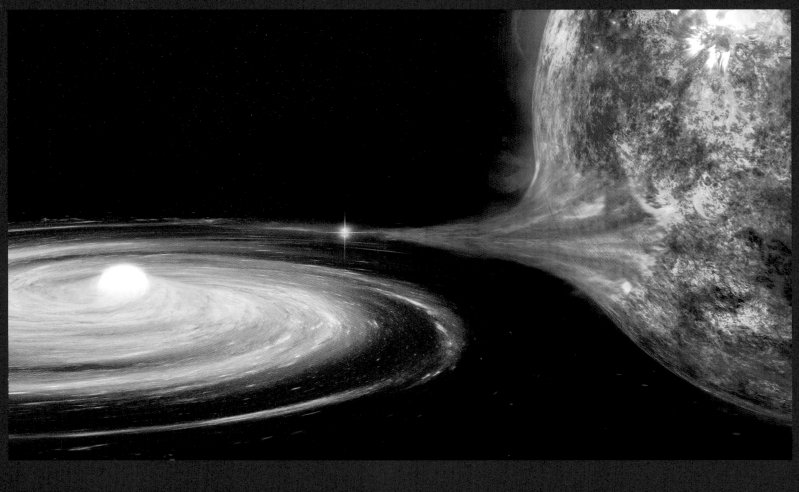

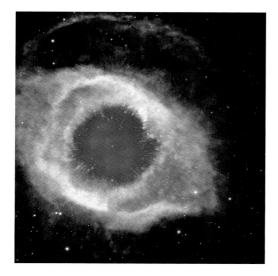

What life if any is out there we do not know, but if for a moment we allow ourselves to dream, to imagine not just life but sentient life living on one of those alien worlds, we soon begin to realise that standing on the surface of one of TRAPPIST's ancient worlds provides a grandstand seat to watch the unfolding story of the Universe, a perspective on time unlike anything us Earthlings can imagine. Starting with the Universe in childhood we'd see entire galaxies evolving, merging and then collapsing as aeons of time pass as if they were days. When you live on a planet that will exist for trillions of years, time takes on a different meaning. The stars in the night sky come and go, and one day, 5 billion years from now, you'll see a distant star flicker and fade. Our sun, a mere flicker in the skies of the TRAPPIST worlds, has come and gone while this star system is still in its earliest infancy.

Our star, the Sun, which has breathed life into our planet and lived for perhaps 10 billion years, is now no more. A ghost of its former self, all that will remain is a burnt-out husk, a white dwarf star, shed of its outer layers and leaving a core of carbon and oxygen to faintly shine with the heat of its long-lost fusion until eventually it will stop emitting any light or heat, becoming a black dwarf in a cold, unchanging universe.

As yet we don't think the Universe is old enough for any black dwarfs to exist, they are purely hypothetical, but we think that one day that will be the fate of every star like our sun.

The arrival of black dwarfs will mark the beginning of the end of starlight. From our vantage point on our TRAPPIST world, the lights of the Universe have slowly been extinguished, the age of the stars is coming to a close, but TRAPPIST-1 will cling on, lingering long into the future.

Ten trillion years in the future even this ancient survivor will enter its last days, slowly cooling until it too finally meets the darkness. We cannot possibly know if TRAPPIST-1 will be the last star in the Universe, but we do know that the last star will be one of its kind. It will be a red dwarf that brings the final curtain down on the age of starlight, and with the passing of the last star the Universe will become dark once again.

This darkness will mark the end of a story which began in a cloud of gas nearly 14 billion years ago. In the time between the darkness that was and the darkness that will be, there were moments of stupendous violence culminating in a cosmos of startling beauty and variety, and ultimately in the one very special star that is our sun. Alone (so far as we know) through billions of years and countless generations of stars that lived and died, our sun was born, one star that breathed life into a planet, into a civilisation powered by starlight and curious to explore the Universe. Our star, and indeed the age of stars, will end, but while we are here, we can at least appreciate what stars created and acknowledge with renewed awe the brilliance that brought light and life into the Universe.

Opposite top: Illustration of a bright X-ray outburst from what might be the fastest-growing white dwarf scientists have ever observed.

Opposite bottom: Brown dwarf stars (in size they are between a giant planet and a small star) spin and glow in visible light.

Above: Spewing out hot gas and ultraviolet radiation, this star (the central white dot) is in the throes of death and forming a white dwarf.

GALAXIES

'No man is an island,
entire of itself,
every man is a piece of the continent,
a part of the main.'
John Donne (1623)

AN ISLAND HOME

Stand on the edge of the ocean, on a shoreline that seems to beat to an endless rhythm, and you could be forgiven for thinking that the beauty and structure that surrounds you is permanent, immovable, eternal. In the fleeting timespan of a human life, we can at first sight become blinded to the dynamism and change that surrounds us. But look again through a longer lens and you will see a very different story. The shorelines of our planet are littered with the evidence of endless change through Earth's history; where there are soaring cliff tops there was once an ocean floor abundant with life, where slabs of granite tumble into the sea there was once the red-hot magma of a long-dead volcano; and where we see a coastline vanishing into the distance we are looking at the jagged edges of ancient lands ripped apart at the seams. Only with the very deepest sense of time do you begin to see the true story that lies behind the beauty of our world, and what applies here on Earth also applies in the far reaches of the cosmos high above our heads.

In this chapter, we are going to explore the beauty and fragility of our galaxy – the Milky Way. Every star that we can see above our heads is part of that galaxy, but from our vantage point here on Earth we have only been able to glimpse this family of stars from afar, as we are too entwined within it to truly understand the immenseness of the larger structure or the depth of its history.

The beauty and spectacle of the Milky Way seems endless, a majestic island universe that has hardly changed since humans first walked the Earth. But the fleeting stability that we have witnessed hides the real nature of the galaxy we call home. Since its birth billions of years ago, shaped by the forces of gravity and evolving through the most dramatic of collisions, the Milky Way has been a restless and often hostile place, and yet it is this perpetual change that is responsible for the creation of our solar system, our home and, ultimately, us. And one day, far into the future, this thirst for transformation is what will destroy the galaxy and everything within it.

The sheer scale of our galaxy is almost impossible to comprehend. It would take 100,000 years travelling at the speed of light to get from one side to the other. To put that into context, the distance between the Earth and the Sun is approximately 150 million kilometres, while to the orbit of the outermost planet, Neptune, it is around 4.5 billion kilometres, and to our nearest neighbouring star is another 40 trillion kilometres. These are distances that pale into insignificance when you realise that the Milky Way is a billion billion kilometres from one side to the other.

Our galaxy, like the estimated other 2 trillion galaxies in the observable universe (2×10^{12}), is a gravitationally bound collection of stars, gas and dust orbiting a galactic centre of gravity. Galaxies come in all different sizes, ranging from the smallest dwarf galaxies made up of a few hundred million stars to the largest-known galaxies that we believe hold as many as 100 trillion stars. Galaxies also come in different shapes: some like ours are spiral in structure, with arms curving out from the centre, while others, known as elliptical galaxies, are smooth and oval in shape. Some have no distinct structure at all, and these chaotic-looking galaxies are the ones that we categorise as irregular.

The majority of these galaxies are arranged into even larger structures that are bound together by the force of gravity's endless hunger for attraction. Dwarf galaxies orbit their larger neighbours, and grand dances take place in the galactic groups, clusters and superclusters that bind together galaxies in ever-increasing numbers. A galaxy group like our own Local Group can consist of anything up to 50 galaxies bound together in a gravitational dance. Galaxy clusters can comprise thousands of galaxies and superclusters, like the Virgo Supercluster in which we find ourselves, which can bring whole groups and clusters together into some of the largest structures in the known Universe.

Opposite: With 185 million years of geological history, the Jurassic Coast in Dorset, England, is a reminder of our evolving natural world.

Top: M100 is a classic example of a grand design spiral galaxy, with prominent and well-defined spiral arms winding from the dense centre.

Middle: NGC 454 is a galaxy pair of a large, red elliptical galaxy and an irregular gas-rich blue galaxy. Both are warping as they grow closer.

Bottom: Lacking distinctive structure, irregular dwarf galaxies like UGC 4459 can appear chaotic, but they are a common galaxy type.

'There are as many galaxies in the Universe as there are stars in the Milky Way. That's 100 billion – ten times greater than the population of people on Earth.'
Payel Das, Astrophysicist, University of Surrey

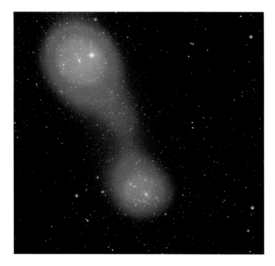

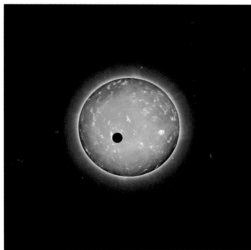

Between all of these trillions of galaxies that span the Universe, there is pretty much nothing. The intergalactic medium is a tenuous gas, a scattering of hydrogen and helium atoms that fills the vast vacuous regions of space where no galaxies exist.

Looking up to the heavens with our most powerful telescopes, we have witnessed a universe full of galaxies, but it's only within our home, the Milky Way, that we have been able to observe the true nature of a galaxy. Inside this gigantic structure are more than 100 billion stars bound together by gravity, all orbiting a supermassive black hole we call Sagittarius A*. From Earth, when we are lucky enough to see it, the Milky Way appears as a band of light arcing across the night sky. It was first speculated that this bright strip of light might be a vast array of distant stars back in the time of the ancient Greek philosopher Democritus, 2,500 years ago. But it wasn't until Galileo lifted his telescope to the night sky 2,000 years after this that we saw for the first time that this structure really was composed of distant individual stars.

Today, we know the Milky Way is filled with an extraordinary array of wonders, structures of endless beauty that chart the birth, life and death of myriad stars. In the vast dust clouds of the Eagle Nebula we have witnessed newly-born stars firing jets of gas into the swirling stellar nurseries. In the Carina Nebula we have watched the mysterious structures of the binary star system Eta Carinae billow into space. Unlike our sun, 80 per cent of the stars in the galaxy are not single stars but multiple star systems like Eta Carinae, a duo of stars which when combined are 5 million times brighter than the Sun and are still getting brighter.

We have also witnessed the remains of long-dead stars, such as in the ghostly halo of Cassiopeia A, a bright remnant of a supernova explosion that ripped through the Milky Way approximately 11,000 years ago but only first appeared in our night sky in the late seventeenth century.

As well as billions of stars, we also now know there are billions of planets in orbit around the vast majority of stars in the Milky Way. One of these planetary systems, Kepler 444, is found in the constellation of Lyra, approximately 119 light years from Earth. Kepler 444 is a triple-star system with a main-sequence star like our sun at its centre and two red dwarf stars spinning around each other and the mother star in a highly distorted 198-year orbit. One of the oldest systems in the Milky Way, it is estimated to date back 11.2 billion years, with five ancient rocky planets in orbit around the main star. These worlds are almost as old as the galaxy itself, planets that have been witness to events long before our own planet came into being.

And yet for all of their wonder, every planet and every star in our galaxy is forever out of reach. In a single lifetime, we could never hope to travel to even the nearest star system, Proxima Centauri – it may be our galactic nextdoor neighbour, but it is over four light years or 40 trillion kilometres away. Yet, despite these vast distances, we have found a way to break free of our limitations and set out on a quest in search of our origins. Science allows us to view the Milky Way from impossible perspectives in both space and time and to tell its story. And as we've allowed our imaginations to wander through the galaxy guided by science, we've discovered that far from being a distant entity, its story is inextricably linked to our own.

Top: A bridge of hot gas connects galaxy clusters Abell 401 (top) and Abell 399 across approximately 10 million light years.

Bottom: A planet transits its mother star in Kepler 444, allowing it to be detected from Earth using transit photometry.

Bottom: Cassiopeia A, imaged one year apart, is a neutron star surrounded by the 325-year-old infrared echoes of its own explosion.

Right: The Pisces-Cetus Supercluster Complex, which is a billion light years across, is home to the Virgo Supercluster, centre.

<u>What is a galaxy?</u>

If you think of the Solar System, we have a star, the Sun, and a whole bunch of planets and asteroids and meteors ... So a galaxy is really just a collection of many of these kinds of systems, replicated several thousands and millions of times over. It is, in a very simple sense, just several hundreds of millions of stars all put together, with a lot of intergalactic gas, primarily in the form of hydrogen. And they're all orbiting under a common gravitational force. What separates the galaxy from just a collection of stars is that galaxies typically tend to have a very, very massive black hole right at the centre, known as a supermassive black hole. They also tend to contain a vast amount of dark matter, which is what holds everything together.

Sownak Bose, Researcher, Center for Astrophysics, Harvard

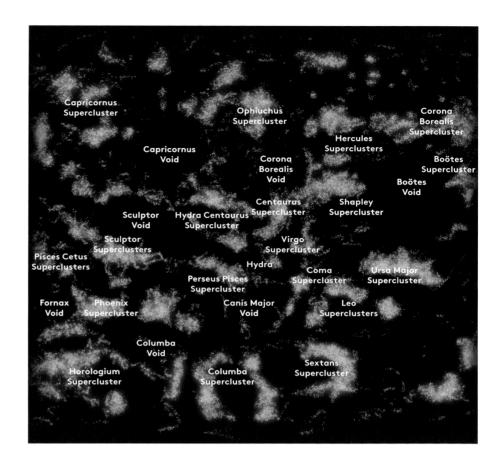

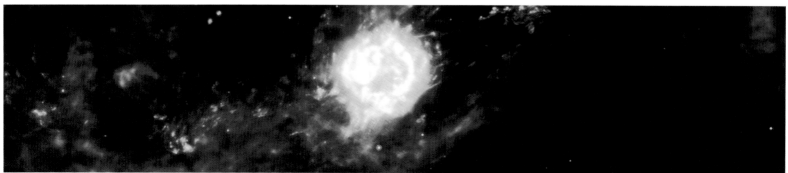

In order to be able to see the Milky Way from our perspective here on Earth, you have to live far away from a city centre. So, if you are ever up in the mountains and it's a pretty clear night sky, you might see this band of stars stretched across it, which corresponds to the disc of the Milky Way. We actually live inside one of these spiral arms of the Milky Way, around two-thirds out from the centre of the galaxy. We as human beings on planet Earth are located somewhere towards the suburbs of the Milky Way. If the viewing conditions are good, we can see some of the surrounding stars that are part of this spiral arm.

The Milky Way is an example of a spiral galaxy. Spiral galaxies are these really beautiful creations of cosmic evolution where you basically have a very dense central portion, which is where a lot of the stellar light is concentrated. So there are thousands and millions of stars constantly right in the centre of our galaxy. And around the centre are a whole bunch of spiral arms, which are the locations where typically a lot of new stars are created. You can almost think of the birth of new solar systems taking place all the time in each of these spiral arms. If you were to weigh up all the stars it contains, you'd find that the stellar content of the Milky Way weighs roughly around 50 billion times the mass of the Sun.

All galaxies have a lot of dark matter as part of their makeup. In the case of the Milky Way, the current estimate is that between 10 and 15 per cent is in the form of ordinary matter, whereas the remainder – around 80 per cent of it, or a little bit more – is in the form of dark matter spread throughout the galaxy. If you imagine this composite system of dark matter and ordinary matter at the very centre of this system, this is where the ordinary matter dominates, whereas at the exterior of this overall composition is where most of the dark matter is. So we are essentially surrounded by this halo of dark matter.

The Local Group
The Milky Way and its neighbours as seen from a location about 700,000 light years from the Sun. The nearest galaxies are the two irregular dwarfs, the Magellanic Clouds. In the distance beyond them is the spiral galaxy M33. There are about 50 members of the Local Group, but not all show up in this perspective. The location of our sun is marked by a red dot. In reality it would be invisibly small from this distance.

NGC 147 & NGC 185
This is a pair of dwarf elliptical companion galaxies to the Andromeda Galaxy. Perspective makes them appear close to our galaxy. NGC 278 is a much more distant elliptical galaxy.

M110 & M32
These are small, elliptical satellite galaxies of Andromeda.

M31 Andromeda
The Andromeda Galaxy is about 2.5 million light years from the Sun.

M110

M32

NGC 147

NGC 185

MILKY WAY

NGC 278

The Sun

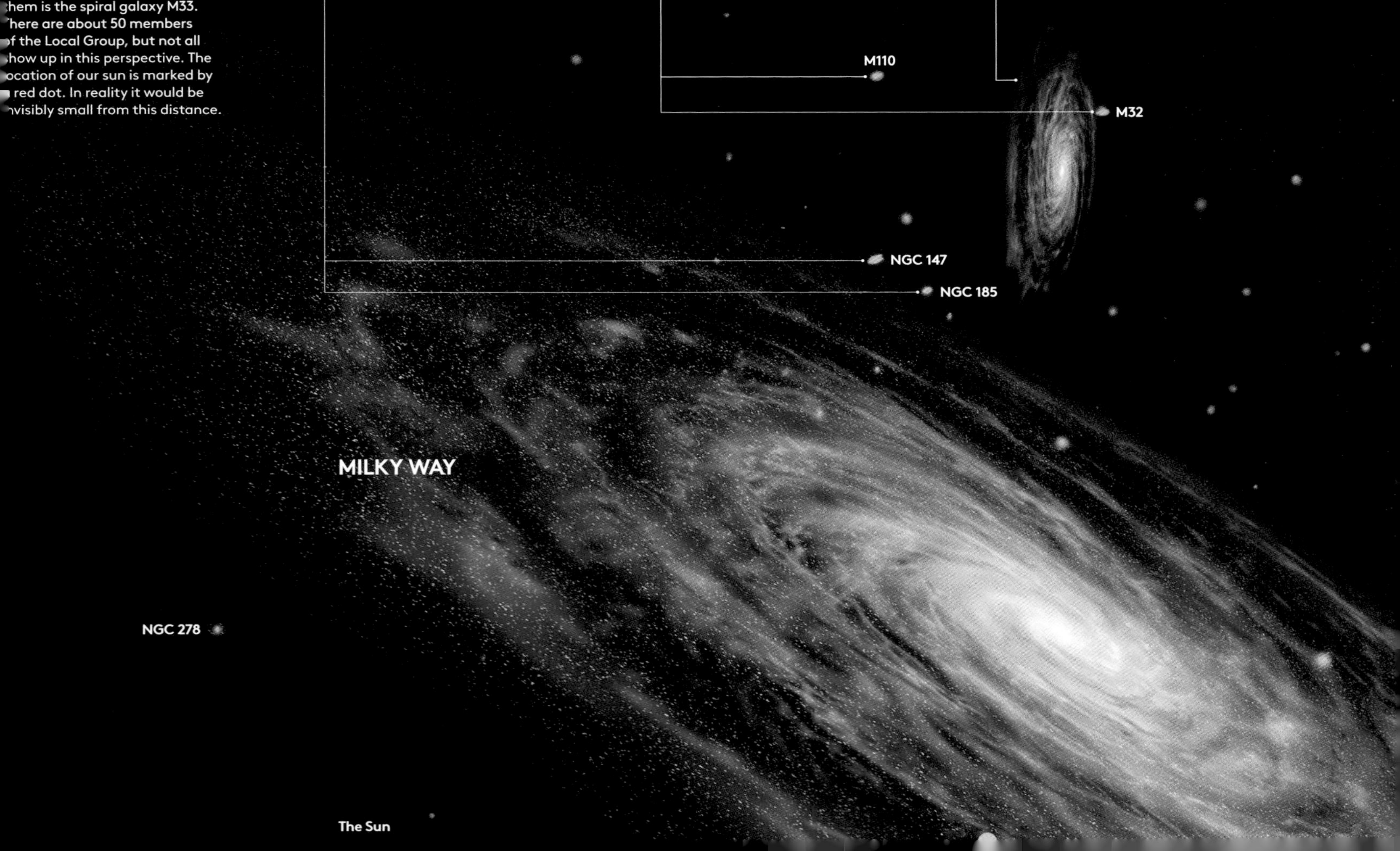

Inside the bulge
Infrared image showing a population of massive stars and complex structures in the hot ionised gas at the centre of the Milky Way.

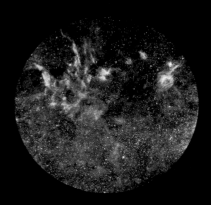

Cosmic filaments
Huge filamentary structures of gas and dust reveal how matter is distributed across the galaxy. This image shows a filament called G49, which contains 80,000 suns' worth of mass.

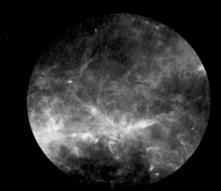

NGC 404
NGC 404 is an elliptical galaxy just past the limits of the Local Group and may not be bound to it gravitationally.

Small Magellanic Cloud

The Magellanic Clouds
These are nearby satellite galaxies of the Milky Way.

M33 Triangulum
At 2.8 million light years from the Sun, the Triangulum Galaxy is the most distant member of the Local Group.

Large Magellanic Cloud

Globular clusters

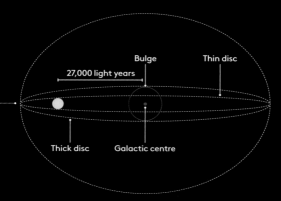

Bulge

Thin disc

27,000 light years

Thick disc

Galactic centre

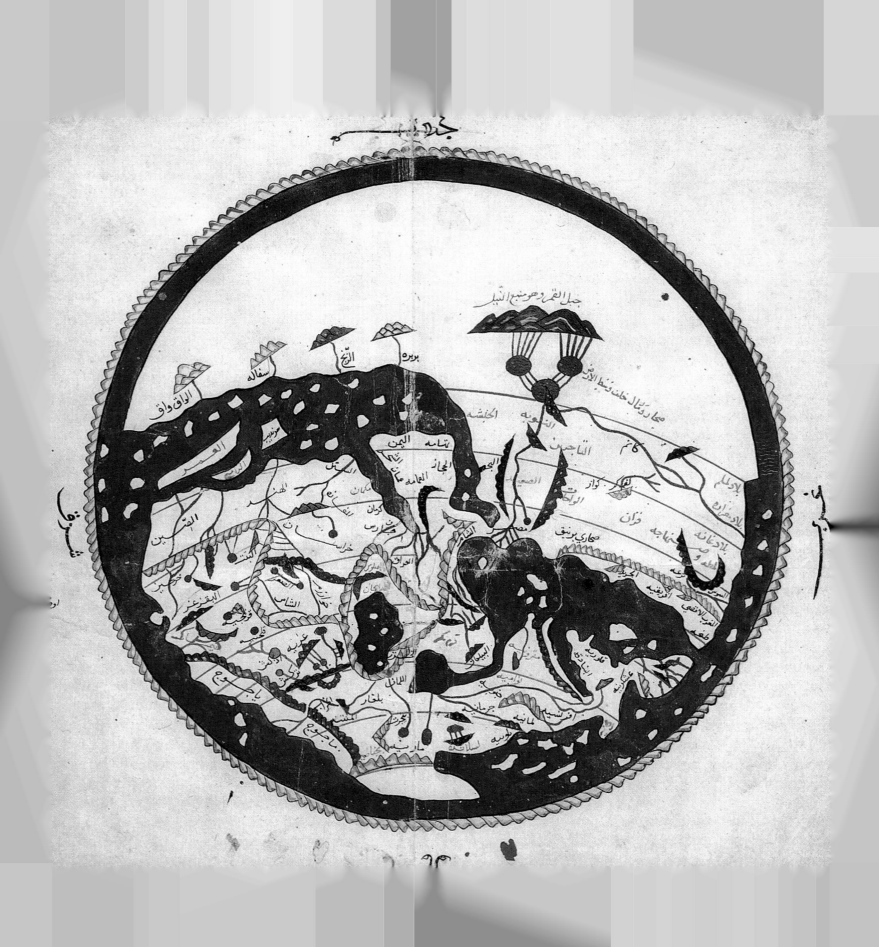

MAPPING THE ISLAND

For as long as humans have had the instinct to explore, the art and science of cartography has never been far away. To create a map by translating the information of a three-dimensional world into a visual form allows us to navigate our way through that world and ultimately gives us the ability to understand how we fit into it. A map can be studied and shared, dispersing the knowledge it contains beyond the experience of those who made it and into the hands of all those who can read its scale and symbols. Maps can also develop, being improved on over years, centuries or millennia as the accuracy of our mapping techniques increase. This is exactly what we have seen over the centuries, as our cartographic skills have developed thanks to advances in technology. From the compass to the sextant, from the hot air balloon to the use of aerial drones, technology has always been the driving force in our ability to map the world around us.

And when we reach back into the past and look for the earliest-known maps ever created by humans, they are not drawings of long-lost shores or river valleys. It seems our distant ancestors' earliest instinct was to record not what lay beneath their feet but what was above their heads.

The Lascaux Caves in southwestern France are famous for the hundreds of wall paintings that adorn the interior chambers and ceilings of this vast cave system. Created over multiple generations, the depictions of the animals that lived in this area have been dated to a time around 17,000 years ago, making them one of the most extensive examples of palaeolithic art we have ever seen. Discovered in 1940 by Marcel Ravidat, a local teenager who found the cave system after his dog fell into a hidden shaft, the cave art has been studied intensely, but it wasn't until 60 years later that archaeologists noticed that the drawings were more than just paintings of animals, they were also maps of the night sky. Hidden within the paintings are images of the constellations, including that of Taurus in the image of a rhinoceros, Leo as a horse, and the depiction of a comet strike within the figure of a dying man. Each is in the exact position that correlates to the formation within the night sky that we know would have sat above these artists' heads at that time.

It seems the instinct to observe the night sky, to question it, track it and ultimately gain an understanding of our place in space and time, has sat within us since deep in our history. A direct line exists from those ancient ancestors scratching their knowledge of the night sky onto the walls of a cave to the twenty-first-century technology we employ to map the stars today. Across thousands of years, we have felt drawn to look up and chart the heavens, but the map we are now in the process of making is of such detail that it's allowed us to finally leave our island behind and look beyond its shorelines.

'We have been connected with the night sky for millennia; there is always some kind of celestial connection that humans have tried to draw between our place here on Earth and the vastness of space.'
Sownak Bose, Researcher, Center for Astrophysics, Harvard

Opposite: Al-Idrisi's 1154 map of the Mediterranean shows the south at the top – an alternative perspective on a familiar image.

Above: The Cueva de las Manos in Argentina, painted c.7300 BCE, is evidence of the human desire to understand our place in the world.

19 December 2013
'Attention pour la décompte finale.
Dix. Neuf. Huit. Sept. Six. Cinq.
Quatre. Trois. Deux. Un. Top.
Décollage!'

A mission to map a billion stars. That was the objective for the billion-dollar telescope that sat on top of a Soyuz rocket on the launch pad of the European Space Agency's space port in Kourou, French Guiana, on 19 December 2013. Named after an ancient Greek goddess, the ancestral mother of all life on Earth, the Gaia Telescope took three weeks to reach its destination, L2 Lagrange Point, around 1.5 million kilometres away from Earth. Settling into its orbit in gravitational equilibrium between the Sun and the Earth, this spacecraft permanently turned its back to the Sun as it deployed a 10-metre-diameter sunshade that shielded the sensitive instrumentation of the telescope from the heat of our star and at the same time powered itself using the solar panels covering its Sun-facing surface. Gaia then went through months of calibration and testing before it began full operations in July 2014, mapping the heavens with a scale and accuracy unlike anything that had gone before.

Gaia is the most accurate space telescope ever built. In fact, it is actually two identical telescopes separated by a 106.5-degree angle that merge into a single path for the light from galaxies' stars. Each of the telescopes captures the light of a star on its 1.49 x 0.54-metre primary mirror before sending it through an array of ten mirrors of different shapes and sizes that help focus and direct the starlight (it travels over 35 metres within the 3.5-metre spacecraft) before it reaches the 1-billion-pixel camera and the spaceship's three primary scientific instruments.

GAIA'S VIEW OF THE GALAXY
Gaia simultaneously observes two fields of view separated by a stable basic angle while slowly rotating at a constant angular rate of 1 degree per minute around a spin axis perpendicular to both fields of view, to image a circle of the sky every 6 hours.

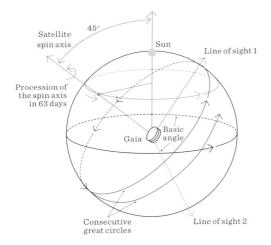

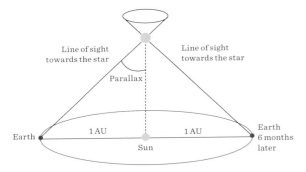

The cosmic cartography that takes place in Gaia's astrometric instrument first measures the position of a star, then combines the repeated measurements of the target star across multiple years to allow for its distance and motion to be calculated as well. In the first five years of Gaia's operation, the telescope was designed to measure each of its target stars at least 70 times, to give an unparalleled set of data to generate accurate coordinates and velocity.

Alongside the astrometric instrument, the photometric detector provides important information on the spectral colour of a star, revealing key properties such as temperature, mass and chemical composition, which in turn allow the age and history of every star in Gaia's gaze to be read. The final instrument is the radial velocity spectrometer, which, by measuring the Doppler shift of its target star, is able to dramatically increase the accuracy of our measurement of the movement of the star when combined with the parallax technique of the astrometric instrument.

Gaia performs all of this delicate observation as it continuously rotates on its axis, using its nitrogen-fuelled micro-propulsion system to make tiny precision adjustments to maintain its fixed position. Observing around 100,000 stars every minute, the spacecraft has already exceeded the limits of its technical specification, examining stars in the magnitude range from 3 to 20 as it slowly builds its map of a billion stars. The information we have gained from Gaia has already transformed our understanding of the Milky Way and our place within it, creating the most precise three-dimensional catalogue of our galaxy. The telescope is now scheduled to stay in operation until at least the end of 2024 (at the time of writing). But Gaia is not just giving us a new understanding of the structure of our galaxy, it is also revealing its life story, its journey through time.

By halfway through its mission, Gaia had already created the most detailed map of our galactic island that's ever been produced. Charting the position of over a billion stars, it has not only created a precise visualisation of our contemporary galaxy, but it has also given us the ability to travel through time. By plotting the velocity and direction of this vast number of stars and combining it with the individual characteristics that Gaia has provided, we can plot how these stars are going to move and where they are going to go, allowing us to fast forward into the future and see what our galaxy will become. And what we can fast forward we can also reverse; contained within the motion of the stars is the story of where they came from, and reversing the direction of the stars allows us to wind back the story of the Milky Way, providing a first glimpse of our galaxy billions of years in the past, long before our own solar system was born, back to a time when the very existence of the galaxy itself was far from certain.

Gaia has created not just the most extraordinary three-dimensional map of the Milky Way but also a four-dimensional map that has in turn created a whole new science: the science of galactic archaeology. And digging into this past has revealed that far from being a sedate island of stars, our galaxy has lived the most dramatic of lives, survived events on a scale that is unimaginable and somehow contrived the conditions for our sun and our planet to nurture life. But as with every story we find in the far reaches of our own history, it could easily have ended so differently. In an endlessly violent and volatile universe, not every young galaxy makes it to adulthood.

Opposite: Paintings in the Lascaux Caves, c.15,000 BCE, are thought by some to be the earliest known examples of star maps.

Above: The distance to a star can be calculated with simple trigonometry comparing the measured parallax angles from Earth.

LAUNCHING GAIA

Professor Gerard Gilmore, Institute of Astronomy, Cambridge University

I was one of the half-dozen people who wrote the original proposal for Gaia back in the early 1990s. I was involved in the team that was managing all the set of compromises and studies that you have to do when you go through the initial engineering match against the science to see what you can actually deliver. And I was there at the launch, 20 years later, at Kourou in French Guiana, a high-tech launch site in the jungle in South America.

The launch was just before sunrise on a beautiful, clear day, it was just perfect. The special thing about Gaia was it was not a standard rocket launch. If you've seen one, the rockets go straight up and in a burst of flame, they disappear in the cloud. All you see is a sort of bright glow for 20 seconds, then you hear somebody giving you the description over the radio. It took four minutes for Gaia. You could see the flame of the rocket and you could see the individual stages popping off, then eventually it disappeared over the horizon.

It went up into a parking orbit for a couple of hours while the technicians just sort of slowly turned things on and off before they got into this critical state where they had to open up the sunshields that are 10 metres across. That was the do-or-die moment for Gaia; it was critical that these opened up to protect the payload from the Sun.

Finally, a text message arrived to let us know of the successful opening. After nearly 20 years of work, it was something of a relief.

Advanced payload technology
Dual telescope concept in a single integrated instrument comprising:
10 mirrors
1 astrometry function
1 photometry function
1 spectrometry function

Permanent data link to Earth
Download operational
8 hours a day at 5Mbits/sec
Five years of data will be 1,000 million million bytes

The largest instrument ever built using ceramics
Structure made of silicon carbide, a material optimised for stability, durability and low mass

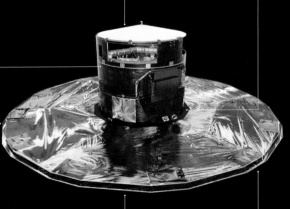

Height: 3 metres
Diameter: 10 metres
(sunshield deployed)

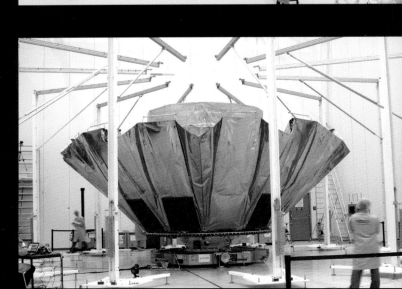

Measuring instruments of unprecedented precision
Photometer with a resolution of 1 billion pixels capable of detecting stars with a luminosity 400,000 times lower than those visible to the naked eye

Thermal insulation
Resistant to temperatures between -170 and 70 degrees Celsius

Extremely high pointing accuracy
Cold-gas micro-propulsion system for fine attitude control and maximum stability

MAPPING HISTORY

To understand where we came from, we first need to understand where we are actually sitting, in terms of the Universe, today. From our small rock, it's difficult enough to imagine that we are part of a solar system, a system of planets, dwarf planets, moons and asteroids orbiting the Sun, but we are, as we know, part of a system so much bigger than this. We are not just part of a solar system, not just part of a galactic system, but in fact part of a multigalactic structure, a vast swirl of galaxies bound together by gravity that we call the Local Group.

The Milky Way is just one of more than 54 galaxies clustered together in our local galactic neighbourhood. Most of these are dwarf galaxies that are clustered around the three giants of the Local Group – the Milky Way, the Triangulum Galaxy and, the biggest of them all, the Andromeda Galaxy, a vast structure that contains over a trillion stars. Just as we orbit the Sun, so our galaxy is in a gravitational dance with Andromeda, orbiting around a gravitational centre that sits between these two islands of stars, a dance that is slowly bringing them closer together.

'With the precision achieved, we can measure the yearly motion of a star in the sky which corresponds to less than the size of a pinhead on the Moon as seen from Earth.'
David Massari, University of Groningen

Beyond this we find ourselves part of an even bigger structure – the Local Group is on the outer edges of a giant supercluster of galaxies known as the Virgo Supercluster. This vast swirling system of gravity-bound galaxies is thought to be over 100 million light years across, with at least 100 galaxy groups, including our own Local Group, bound together within it. And yet this is not the end; every time we seem to reach the largest structure, we find that another home emerges from the darkness.

So before we take one more step out, let's recap: the Solar System is part of the Milky Way Galaxy, which is part of the Local Group of galaxies, which is part of the Virgo Supercluster, which we now know is part of the Laniakea Supercluster, a structure that we believe is home to approximately 100,000 galaxies. And yet even this vast collection of galaxies that stretches out across 520 million light years is not the end of our journey, because Laniakea is part of an incredibly long chain of galaxies that we now call the Pisces-Cetus Supercluster Complex. A billion light years in length and 150 million light years wide, Pisces-Cetus is one of the largest structures we have ever observed within the Universe, with a total mass of a quintillion (that's one with 18 zeros after it) suns within it.

Pisces-Cetus is more than just a grand collection of galaxies. It's a structure known as a galaxy filament – and it's not the only one. We have discovered a series of these massive threads spiralling across the Universe, ranging in size from a couple of hundred million light years to well over a billion light years in length. Stretching across unimaginable distances, these galaxy filaments form a web between which there is nothing, just voids in the Universe. What we are seeing, in the glimpses that we've caught of this giant web, is that on the very largest scale the Universe has structure – galaxies are not just scattered randomly across the Universe, there is order in the vastness.

It all began in the darkness, 12.8 billion years ago, with a web of dark matter strung across the infant Universe, a structure that hid its secret well. This was the cosmic web, a vast network of strands and filaments stretching out for billions of light years. This enormous thread of dark matter provided a gravitational magnet around which immense clouds of gas and dust could begin to gather and slowly condense into a structure large enough not just to form stars but a galaxy of 400 million stars – the Milky Way.

In the earliest part of its history, as this massive cloud of gas and dust gave life to millions of stars, it began to collapse into a disc-like structure, the beginnings of the galactic shape we see today. But in those nascent years, there were no guarantees that a stable assembly would form. Our galaxy was a fragile embryo surrounded by hundreds of other embryonic galaxies, all close together, colliding within this fragment of the cosmic web, and all struggling to emerge from the primordial chaos. As it set out through the Universe, the young Milky Way was hemmed in on all sides by galaxies both large and small, in an environment where some galaxies were predators and some were prey. In this endless gravitational war, survival meant devouring any other galaxy that came into close proximity.

Left: The Milky Way arch dominates the sky over Akaroa, New Zealand.

Opposite: Spirals form in galaxies because they balance out the gravitational forces of the stars.

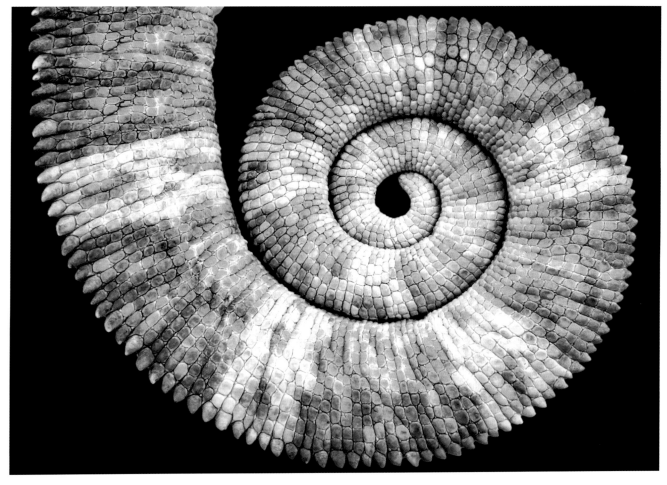

Right: Spirals occur commonly in nature, in the growth of shells, plants and animals, and seen here in the curl of a chameleon's tail.

Right: Galaxies form in response to gravity from multiple stars in the same way that whirlpools are formed by pressure from opposing currents.

HIDDEN
HISTORY

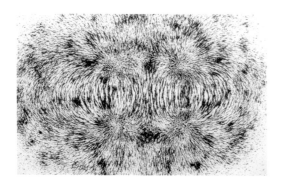

In just the space of a few thousand years of human curiosity, we have been able to look out from Earth and gradually piece together our galaxy's structure – with its central bulge, its distinctive spiral arms and its thin, wispy halo. Looking out into the enormous starry arc stretching across the night sky, we have gradually been able to explore our place in the Milky Way, but we have struggled to look too far back into its violent past. Until very recently, the detail of the early history of the Milky Way has been lost to us in the billions of years that have elapsed since those early violent times. But, as is often the case, the Universe leaves traces of its past, and with the right technology, clues that have remained hidden from our view can suddenly come into plain sight – you just have to know where to look. And that's where Gaia has transformed our ability to wind back the clock on the Milky Way.

On 25 April 2018, the Gaia Space Telescope team released the second tranche of data from its ongoing mission. DR2, as it was called, was based on 22 months of observation by the spacecraft between the summers of 2014 and 2016. Detailing the position, movement and parallaxes of over 1.3 billion stars, it was a treasure trove of data that literally had researchers working through the night to see what secrets lay within this new map of our home island of stars.

One of the scientists racing to get their hands on the DR2 data was Professor Amenia Helmi and her team from the Kapteyn Astronomical Institute of the University of Groningen. Having spent years trying to unpick the origin of the galactic halo that circles the Milky Way, Helmi and her team were desperate to sift through the data and be the first to see any new finds that were leaping out of it. The advancement of science can often be deadly slow and endlessly collaborative, but when a new map of such power is falling into the inboxes of every researcher on the planet at exactly the same time, it can, unsurprisingly, become intensely competitive as different teams race to see what gems lie among the box of treasure.

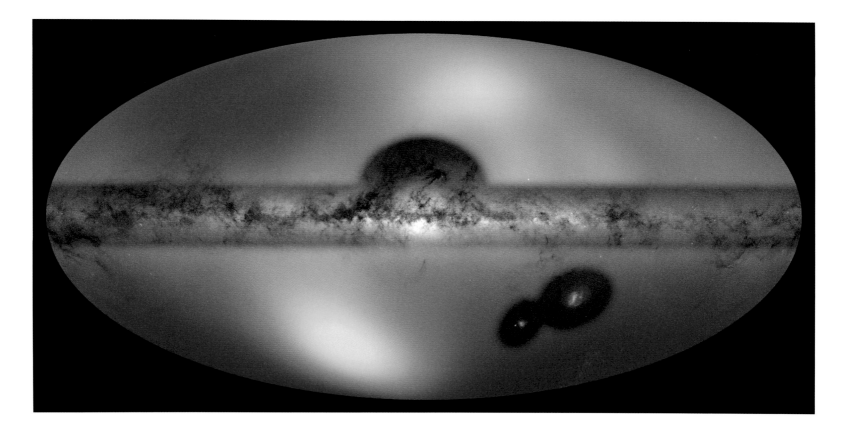

Opposite top: Iron filings are used in an experiment to illustrate the magnetic field lines on Earth.

Opposite bottom: The Milky Way and the two Magellanic Clouds against the galactic halo. Stellar density is higher in lighter regions.

Below: Light emanating from polarised dust. The texture indicates the galactic magnetic field lines.

Bottom: The towers of fiery colours in Planck's map of the Milky Way are dust particles that have been polarised by reflected light.

Within hours of the data landing, Helmi and her team were following their first lead. Gaia's latest map had revealed a collection of 30,000 stars that were moving in the oddest of directions. Just like in our solar system, the Milky Way is full of star systems that are all moving in the same direction within the same flat orbital disc around the galactic centre. But this collection of 30,000 stars was doing none of this. Instead, they seemed to be moving against the flow of the galaxy, orbiting backwards, and not only that, they were travelling in orbits that were carrying them out of the flat galactic plane. This was no anomaly; it was the first direct evidence we had of a moment of enormous violence that would transform the character of our early galaxy forever. Rewind the tape and these strange stars turn into more than just a galactic oddity, they lead us back to a long-hidden moment in the Milky Way's history, that crucial moment when our infant galaxy collided with a long-lost neighbour.

Somewhere in the distant past, in the first few billion years after the Milky Way came into existence, it found itself in a dance with another galaxy. A neighbouring galaxy that had for millions, if not billions, of years circled us just like a planet orbits a star. We have named that long-lost galaxy Gaia-Enceladus (after the giant, Enceladus, who descended from Gaia, in Greek mythology) and we believe that this small dwarf galaxy crossed a threshold around 10 billion years ago and changed the destiny of our Milky Way forever. As the smaller galaxy came into touching distance of our galaxy, waves of gravity came rippling in its wake, millions of stars were hurled into disarray while a titanic galactic battle took place. For all of the chaos that ensued, there would be only one winner – in a battle of this scale size really does matter, and the gravitational might of our galaxy meant that it would be our home island that would grab the incoming galaxy and hold the stars of Gaia-Enceladus tight within its grip.

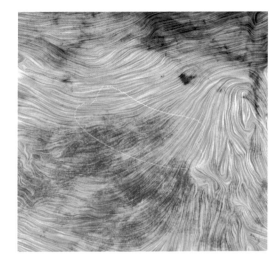

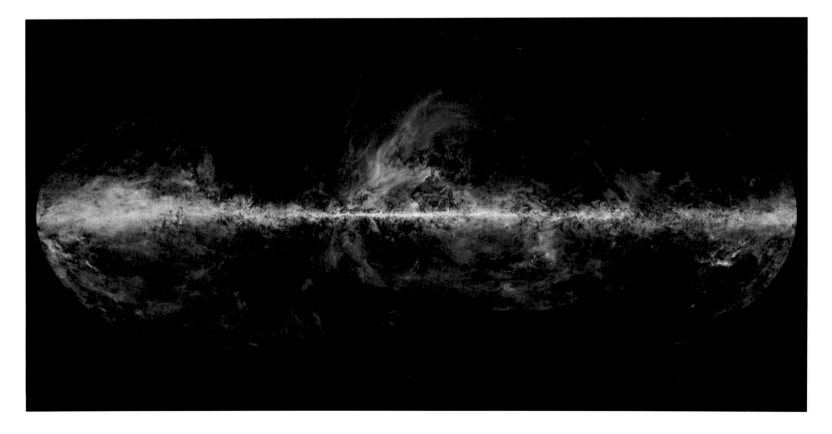

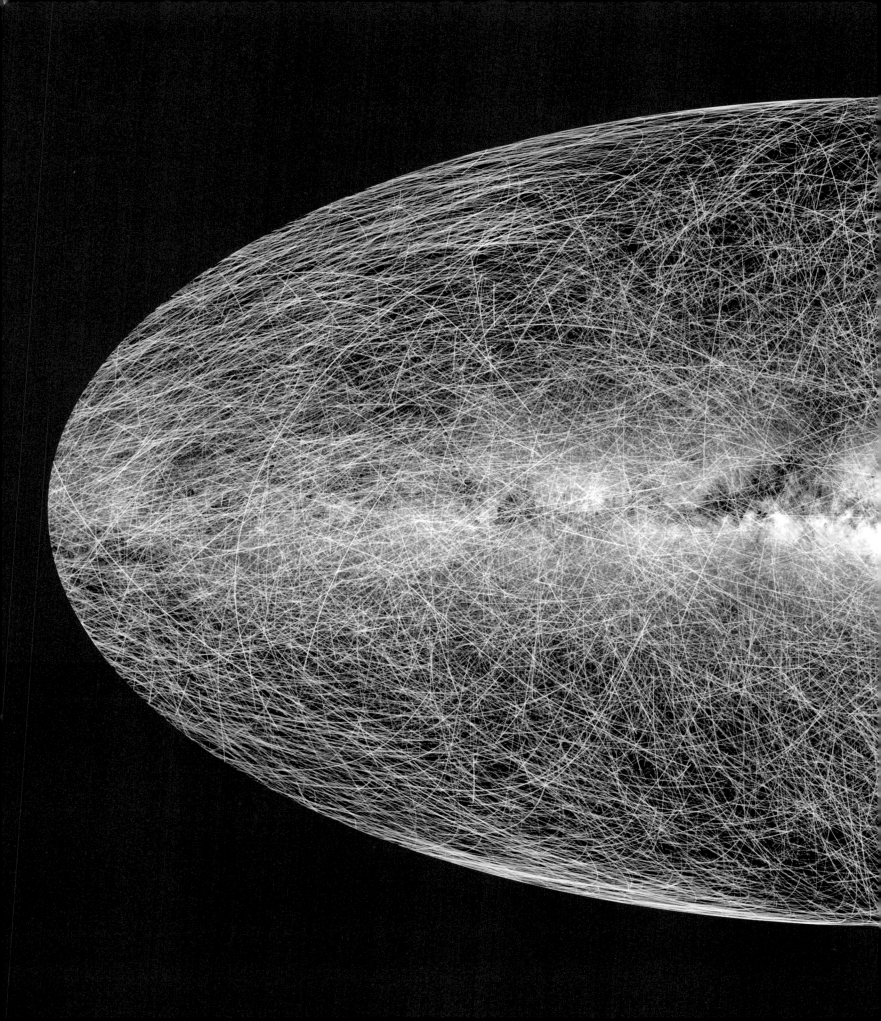

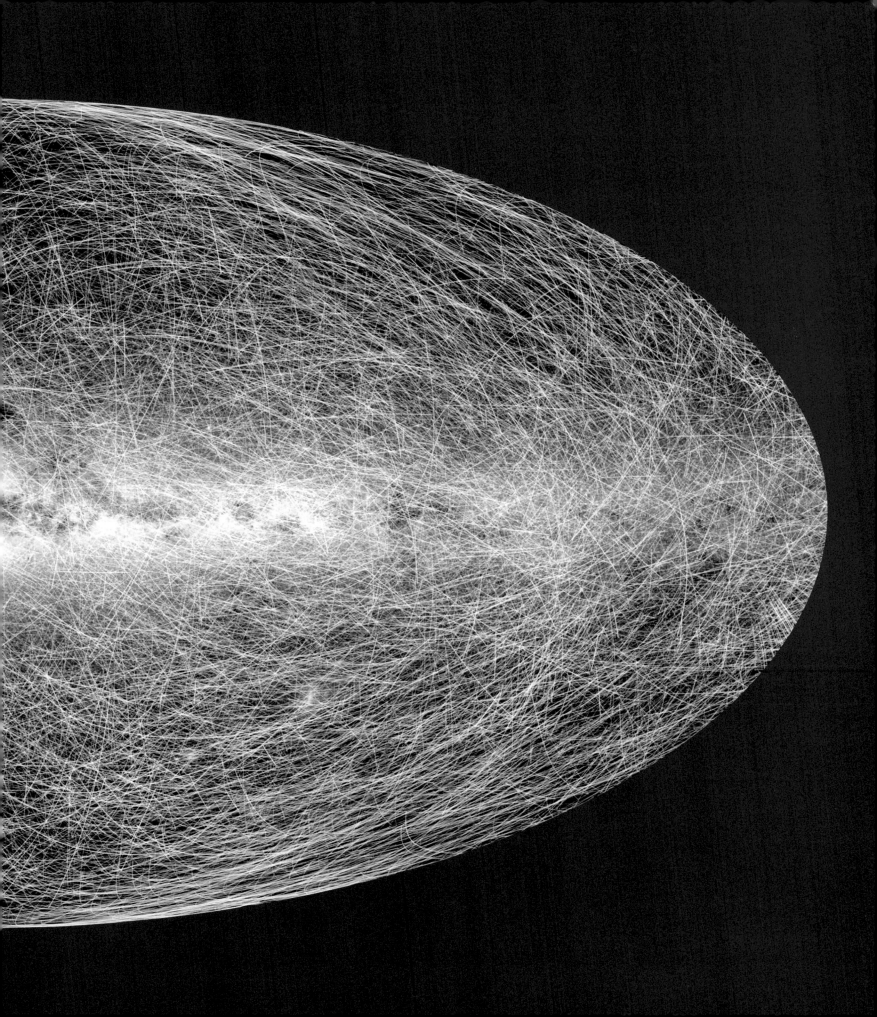

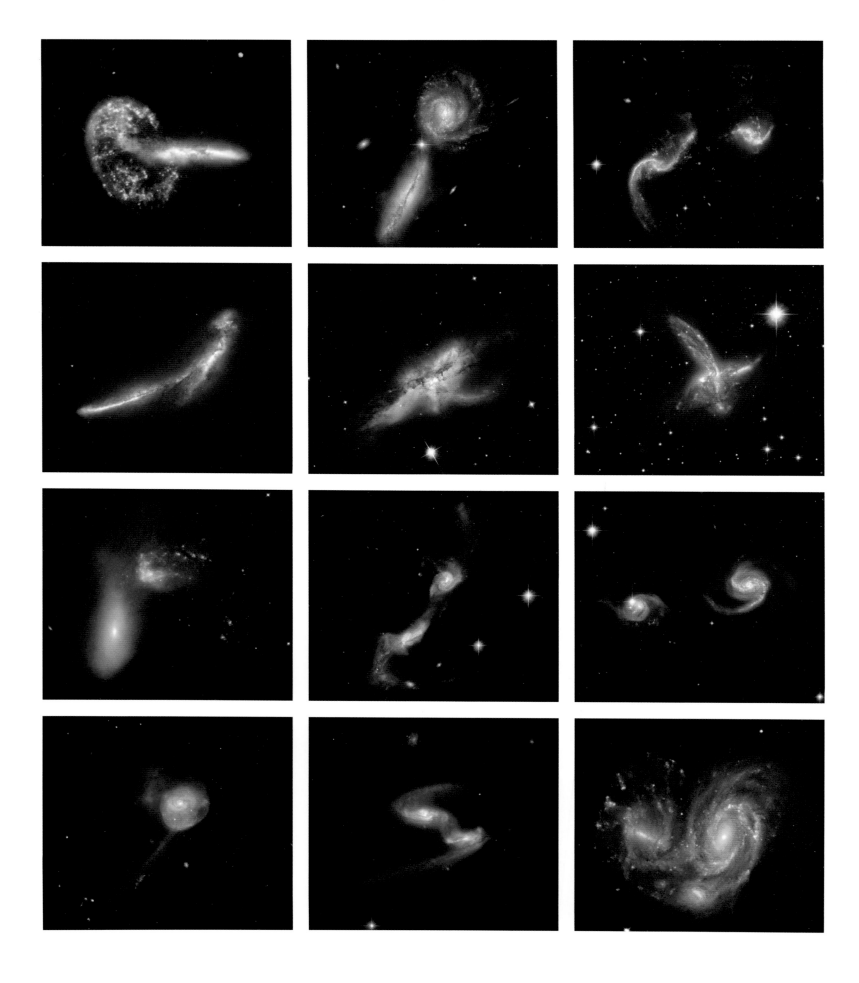

Gaia-Enceladus collision

Galaxies colliding is a very interesting process; it might sound like a very destructive sequence of events, but if you actually run a cosmological simulation of two galaxies colliding, it's really quite beautiful and actually a pretty formative event. The way in which the mutual gravity of these two galaxies interacts with each other causes one to start spiralling around the other, before it comes back. It's like two objects performing a sort of celestial ballet around each other, until eventually one day friction is able to bring the smaller of the two objects crashing in towards the centre of the galaxy.

When galaxies interact with one another and collide, what typically happens is that you actually get a big burst of star formation occurring, because you are essentially bringing a new source of star-forming fuel into the Milky Way. In terms of Gaia-Enceladus, it brought in with it its gaseous content, then as it fell in and plunged into the Milky Way, this gaseous content gets compressed and actually has high-enough densities that are needed to start the formation of new stars. But this process can also lead to disturbances in the existing configuration of stars in the Milky Way. So, for example, if a galaxy plunging into the Milky Way gets close enough to the disc, you can actually start creating warps in the distribution of stars and oscillations in the motions of stars in or around the disc. Which means both events really don't go completely unnoticed.

The Milky Way has experienced a large number of these kinds of emerging events, with Gaia-Enceladus being one example, and one of the big implications of this is that there is a sizeable fraction of the stars in our Milky Way that were actually not part of the system originally, that came in from a neighbouring galaxy. So, the stellar content that we see around us today is really the sum total of intrinsic star formation within the Milky Way, but also the donations from all of the other neighbouring galaxies that came in. Thus, each of these stars actually maintain, in a weird sort of way, a fingerprint relating to their origins, which helps us decode the kinds of events that the Milky Way underwent in its past.
Sownak Bose, Researcher, Center for Astrophysics, Harvard

The Milky Way had collided with a galaxy that was a quarter of its size and swallowed it whole, leaving behind a trail of nomadic stars that would wander unnoticed on a crooked course through our galaxy for billions of years. Only with a map, drawn up using the knowledge of thousands of years of human exploration, could we look beyond the myopia of our own position in the Milky Way and see these stars for what they really are – alien stars, invaders from another island. (Combining the data from Gaia with ground-based observations from the Sloan Digital Sky Survey in New Mexico, we have been able to add even more credence to this theory, by observing that every one of these 30,000 stars has a similar chemical composition.)

Today, we see the remnants of this long-lost galaxy in the off-kilter trail of those 30,000 stars as well as the presence of at least eight globular clusters of stars that roam our galaxy. One of these clusters, NGC 2808 is a large cluster of over a million stars, composed of three generations of stars all born within 200 million years of each other. Such an unusual cluster of stars points perhaps to this being the remains of the galactic core of that long-lost galaxy that collided with ours so long ago.

While we are now able to peer at the aftermath of that crash and see the wreckage continue to reverberate through the Milky Way, exactly what the consequence of this collision had on the evolution of the Milky Way is still being explored. When we think about galaxies colliding, it brings to mind images of Hollywood disaster films, of stars colliding and getting ripped apart. But as exciting as that image might be, it's not what happens at all. The distances between stars are so immense, there is almost no chance of any of them actually colliding. Instead, when galaxies interact, the stars get scattered, the shape of the galaxy may change but nothing gets destroyed and, in fact, we think that rather than being destructive, galactic collisions can often be the engines of creation. And this is what we now think may have been the result of the Gaia-Enceladus collision.

One theory that has growing support suggests that Gaia-Enceladus was instrumental in changing the fundamental structure of the Milky Way. Today, the disc at the centre of our galaxy is comprised of two components – a thin inner disc full of young stars and surrounded by vast amounts of the gas and dust needed to create them, and a thicker disc full of older stars that sandwiches the thin disc on all sides. Exactly how this disc structure evolved remains one of the great mysteries of the Milky Way's evolution, but the evidence of the Gaia-Enceladus collision suggests that the injection of huge amounts of energy into the galactic disc may have been the trigger for its expansion. As the two galaxies collided, the stars within the thin inner disc were thrown into chaos and triggered the creation of the thick outer disc. Over time, all of the gas and dust would have then settled back into the galactic place, reforming the thin disc and triggering a new era of star formation.

Our galaxy had survived, its disc swollen in thickness by the titanic impact, and this was now a galaxy emboldened by the injection of new stars and invigorated by the shock of energy reverberating through its deepest structures. In the process of one galaxy's extinction, our galaxy had been rejuvenated, triggering a new era of star birth, but for all of the abundance one star was still missing. Our sun would not be born for another 5 billion years, at a time when the galaxy was relatively quiet, until something else turned up and disrupted the equilibrium.

Previous page: Created from Gaia's data, this map shows the trails of 40,000 stars as they will move across the sky for the next 400,000 years.

Opposite: This collection of images illustrates the remarkable variety of intricate structures that can form when galaxies collide.

LOOKING UP AT THE HEAVENS

In the first months of 1744, a 13-year-old boy by the name of Charles Messier looked up to the heavens in the skies of north-eastern France and witnessed one of the great celestial shows of the millennia. C/1743 X1, otherwise known as the Great Comet of 1744, was a spectacular sight not just in the night sky but also during daylight hours, due to its incredible brightness. But it wasn't just its brightness that marked this comet as out of the ordinary. Many comets have extraordinarily beautiful tails – or coma – that are produced by a stream of dust and gas trailing behind a comet's nucleus. Very occasionally a comet can have two tails, but in the case of C/1743 X1, as it disappeared below the horizon it left behind a fan of six separate tails extending up into the sky like a grand cosmic plumage. For anyone lucky enough to witness the phenomenon it would have been an extraordinary sight, but for the young Charles Messier it was the moment that would inspire his life's work and turn him towards the study of the stars.

Look at any map of the night sky today and you will see the work of Messier. A professional comet hunter, perhaps in endless pursuit of the beauty he had seen as a young boy, Messier would spend his professional life chasing the detection of comets across the night sky. And as any comet hunter knows, there are plenty of false leads there; stumbling across other diffuse fixed objects that could be mistaken for comets is part of the everyday frustration of such an endeavour. Using his 4-inch telescope in central Paris (the light pollution was less of an issue back then), Messier began cataloguing these distracting objects to make sure he could quickly dismiss them in his nightly search for comets. Creating the first version of the Messier Catalogue in 1774, with 45 objects, it grew to eventually include 110 objects, M1–M110, which could be quickly dismissed by any frustrated comet hunters. Despite the primary purpose of the list being to prevent comet hunters becoming distracted, further investigation of his objects in the coming decades revealed Messier's catalogue to be full of interesting astronomical objects in their own right. Today, we know that the 110 objects detailed by Messier include 39 galaxies, 55 star clusters and 11 nebulae, including four planetary nebulae.

One of these comet-like objects, Messier 54, was first observed in the skies above Paris by Charles and his assistant, Pierre Méchain, in 1778, and for the next two and a quarter centuries it was assumed to be one of the many globular star clusters that can be found within our own galaxy. These spherical collections of stars are relatively common within the Milky Way – at least 150 have been discovered so far, with the vast majority found in a halo around our galaxy's core. The significance of M54 could not have been understood at the time of Messier's first observation, but what we now know is that what he was actually looking at was not a star cluster in our own Milky Way but a dense, faint group of stars from another galaxy. In 1994, M54 was discovered to be part of a small galaxy over 90,000 light years away called the Sagittarius Dwarf, which is in orbit around our own. In the space of just over two centuries, our knowledge of this faint group of stars whose discovery began with the inspiration of a six-tailed comet, has gone from comet-hunting distraction to a central role in the story of our galaxy, its evolution and the eventual arrival of the star system that we call home.

Above: The comet Neowise with two tails: the dust tail (white) and ion tail (blue), which first became visible to the naked eye in July 2020.

Opposite top left: The Messier Catalogue, first compiled by Charles Messier in 1774, lists 110 astronomical objects that are not comets.

Opposite top right and bottom: Contemporary illustrations depicting the arrival of the Great Comet of 1744.

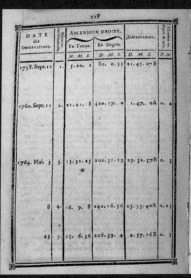

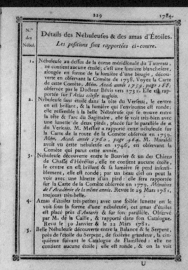

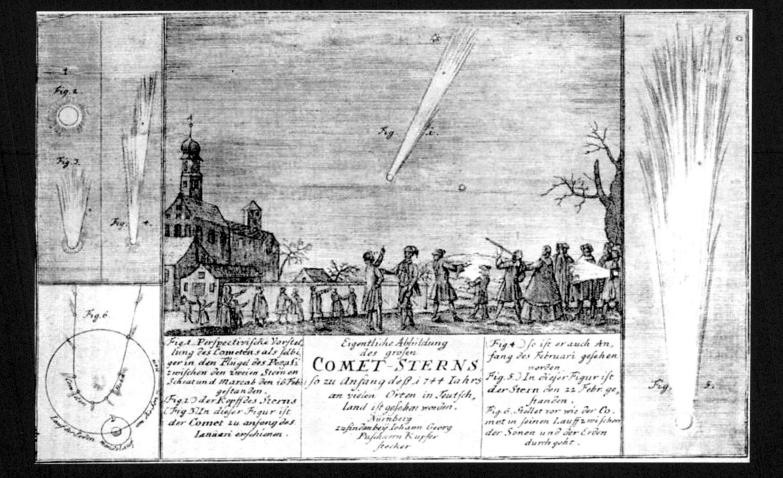

SPOTTING SAGITTARIUS DWARF
The globular cluster M54 is one of the brightest in the sky, but
in 1994, it was found to actually belong to the Sagittarius Dwarf
galaxy, not the Milky Way. It may even be that galaxy's core.

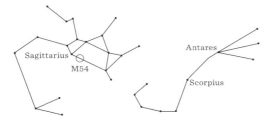

When we talk about galaxies outside of our own, we most often leap to the grandeur of the Andromeda Galaxy (also known as Messier 31), which is the nearest major galaxy to ours at a distance of 2.5 million light years. With a trillion stars within its structure, Andromeda's significance in our neighbourhood cannot be underestimated (more of that later), but much nearer to home there is an abundance of galaxies that are much less impressive but no less influential.

Intimately connected to the Milky Way are at least 59 other satellite galaxies that are bound to us by gravity, many of them orbiting hundreds of thousands of light years away from the centre of our galaxy. We've known about some of these satellite galaxies for thousands of years, when our ancestors looked up and saw the large and small Magellanic Clouds shining in the night sky, the only dwarf galaxies visible to the naked eye, without knowing they were peering at structures containing at least several hundred million stars.

Plotting the trajectory of these galaxies is not easy – we still don't know if the Magellanic Clouds are actually orbiting us or are just galactic travellers passing through. But what we do now know is that of all the galaxies confirmed to be in orbit around the Milky Way, the Sagittarius Dwarf galaxy is by far the largest. With a diameter stretching approximately 10,000 light years across, it's about a twentieth of the diameter of the Milky Way and is currently in orbit 70,000 light years from Earth. Travelling in a vast spiralling path, it loops around the Milky Way on an orbit that takes it over the galactic poles roughly every 600 million years. With its four globular clusters – including the brightest of them all, M54 – we've had our eyes on this galaxy for centuries, but only in the last few years as we've turned the gaze of Gaia towards it have we begun to understand just how important a role this galaxy might have played in the story of our sun, our planet and, ultimately, us.

Right: The bright lights of Messier 54 globular cluster, photographed by Hubble Space Telescope.

Opposite: Collisions between Sagittarius and the Milky Way triggered major star formation episodes, possibly creating the Solar System.

'It is wonderful to think that Gaia and future telescopes such as the JWST will probably be able to pinpoint mounting evidence that Earth was formed right after the collision between the Milky Way and the dwarf galaxy, Sagittarius.'
Rana Ezzeddine, Astronomer, University of Florida

It's not easy being a satellite galaxy. At just 70,000 light years from Earth, the Sagittarius Dwarf skims precariously close to the Milky Way in its orbit, and in this gravitational dance size is definitely not on its side. Over billions of years the heavyweight Milky Way has taken endless bites out of this fragile stellar bundle of stars, using its gravitational might to stretch and strain the dwarf galaxy and strip it of millions of stars. Today, after billions of years, the evidence of this galactic bullying can be seen not just in the diminished remains of Sagittarius Dwarf but in a ghostly stream of stars that trails behind it. On the most distant shores of the Milky Way, this stretches out so far behind its galaxy that it actually now completely encircles the Milky Way. A smoke trail of stars that is only now with the help of Gaia beginning to reveal the intimate dance that has played out on this grandest of scales.

By mapping the Sagittarius stream of stars and plotting this orbital smoke trail in intricate detail, we've been able to rewind the story of Sagittarius Dwarf and reveal the violent exchanges that have punctuated its journey around the Milky Way. Evidence has been uncovered to suggest that this dwarf galaxy has strayed too close to us on three separate occasions. First, about 5 or 6 billion years ago, then about 2 billion years ago, and finally about a billion years ago, Sagittarius Dwarf has collided with the Milky Way, exposing the galaxy to extreme gravitational forces that have ripped it apart, leaving the wreckage of these events of intense violence stretching back in the wake of its orbital trail. Each time the wounded galaxy continued onwards, but each time it has become less of itself, a dying galaxy that will one day succumb to the inevitable and give its entire self to be consumed by the Milky Way. But that is not all Gaia has revealed. As we've pinpointed the moments of collision between Sagittarius Dwarf and the Milky Way, we've also been able to look at the consequences of those collisions on the rest of our galaxy.

8 BILLION YEARS AGO   5.7 BILLION YEARS AGO   3 BILLION YEARS AGO
first Sagittarius passage

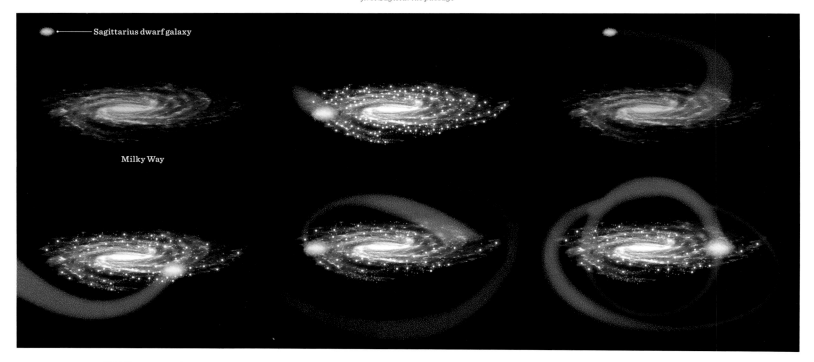

1.9 BILLION YEARS AGO   1 BILLION YEARS AGO   CURRENT SITUATION
second Sagittarius passage   *third Sagittarius passage*

Influence of Sagittarius

The Sagittarius Galaxy was discovered by a student and myself in the 1990s and it's in the late stages of its merger with the Milky Way. Most of the Sagittarius Galaxy is spread out, with streams in front and out the back like giant comet tails wrapping around the entire sky, going out for maybe 100,000 light years away. So most of the Sagittarius Dwarf galaxy and stars are now outside the body of Sagittarius. We could see these, so we knew they were there, but it wasn't possible to understand how they got there and how it worked. Now, with Gaia, we have the motions of these stars, so we can see what direction they're moving in, which ones are going fast, which ones are going slow. We can also see which are going sideways. And sideways comes from the fact that our whole Milky Way is sloshing around. When you consider that the Milky Way itself is live and it's moving around relative to the dark matter, then you can make sense of Sagittarius's tails. And for the first time ever, it's been possible to say, ahh, this is what happened.

When the Sagittarius Galaxy orbited the Milky Way, it came foolishly rather far in – not much farther away from the centre than we are. And this means that it ventured deep into the dark matter and the tides and it got shredded apart. So, this galaxy itself started off as a little galaxy, then it got stretched into these two great long streams, rapidly wrapping around the tail. Those streams were stars, the stars that used to be in the Sagittarius Dwarf, but they're also streams of dark matter that were ripped off Sagittarius and spread around. The timescales and the scale here are so big that these things have only had four or five chances to go around. You can still see them as streams in the Gaia maps, but come back in another 20 billion years and they will all be wiped out and it'll just be background noise somewhere.

But most critically, with Gaia we can see how they're moving. We can see that twisting and more twisting is the key to understanding. We know that when Sagittarius comes close to the disc of the Milky Way, it significantly distorts it. When it goes through the disc it punches a hole in it and the stars get put in these particular spiral patterns, and they stay there, so we can see stars near us that are in those spiral patterns. Sagittarius obviously significantly affected the patch of the galaxy the Milky Way is in. Now we are investigating the possibility that the burst of star formation that occurred when Sagittarius crashed into the disc of the Milky Way created possibly even our own sun.
Professor Gerard Gilmore, Institute of Astronomy, Cambridge University

And when we wind the clock back with this broader view, we begin to see a very different story. Gaia has given us an unparalleled ability to watch how the stars in the galaxy have formed and developed over time, which means we can identify the periods in its history when the Milky Way has been quiet and the periods of sudden change when star formation increases rapidly. What the Gaia data has shown is that over the last 6 billion years there have been three periods when star formation has peaked, 1 billion years ago, 1.9 billion years ago and 5.7 billion years ago, times that align unerringly with the predicted collisions between the rogue dwarf galaxy and the Milky Way. It seems that not only has the Milky Way changed the structure of Sagittarius Dwarf, ripping it apart and spinning its stars into a stream, but Sagittarius Dwarf has repeatedly stirred up the fertility of our galaxy, igniting periods of intense star formation as it has passed through the disc of the Milky Way.

Slowly, the story of our galaxy is emerging. In its infancy came the violence of Gaia-Enceladus, a galactic collision that reshaped the structure of our galaxy forever. Then came the calm, a period of equilibrium when steady star formation and an absence of collisions left the Milky Way in stasis. Waiting for the next great epoch, it would be almost 3 billion years before change would come again. Circling in the darkness, on its grandest of orbits, Sagittarius Dwarf would fall inwards and break the silence. The impact of this collision reverberated throughout the galaxy, sending shockwaves rumbling through the silent and still ponds of gas and dust that had lain undisturbed for eons. Deep within our galaxy, the ingredients of star birth had been stirred.

Today, with the help of our most powerful telescopes we have been able to witness the birth of new stars in our galaxy. Closing in on regions of star birth like the Eagle Nebula (also known as M16), they have allowed us a privileged view of the structure of stellar nurseries. Perhaps most famous of all is the image that has become known as the Pillars of Creation. First photographed in 1995, it revealed vast towers of gas and dust, 30 trillion kilometres high, a thousand times the diameter of the entire Solar System. The peaks of these billowing clouds are illuminated from within as the clouds give birth to new stars, and the dust around them is eroded by the light from hot new stars nearby.

For all of their beauty, these pillars of new stellar life are composed primarily of a single ingredient, molecular hydrogen – the lifeblood of our galaxy. The most common element in the Universe, it is only when it forms into dense clouds of dust and gas that its potential comes to life, and that's where the shockwaves of an impact like that of Sagittarius Dwarf can be so massive.

Opposite: The Eagle Nebula glows as the hydrogen it contains is ionised by radiation from the hot young stars forming within it.

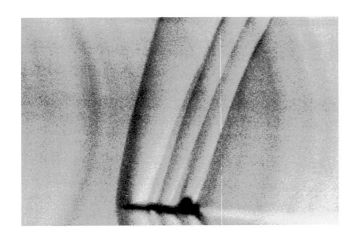

Left: Shockwaves around a T-38 jet aircraft flying at Mach 1.1 (1.1 times the speed of sound).

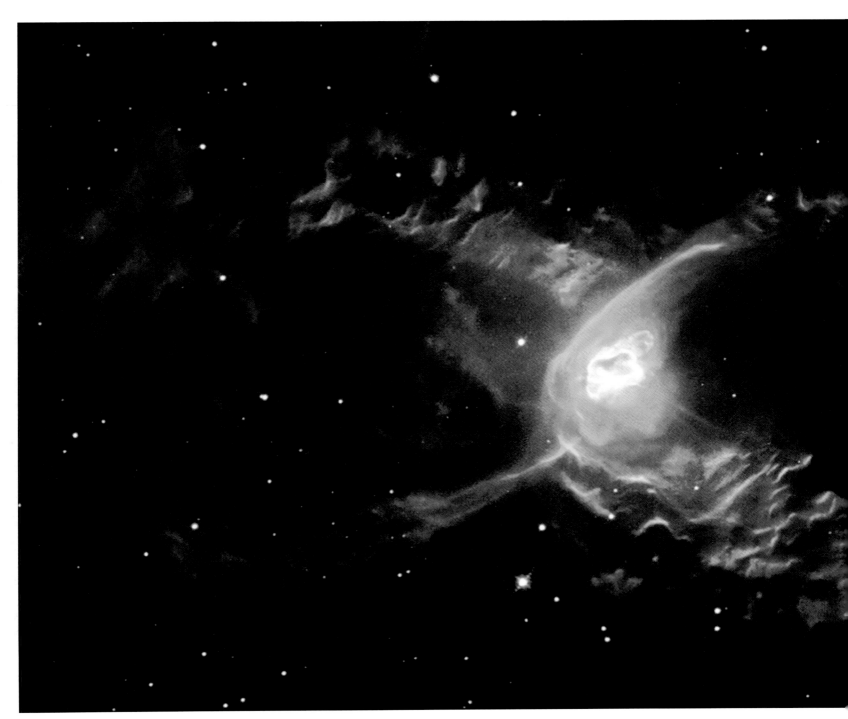

HOW SHOCKWAVES SHAPE NEBULAE

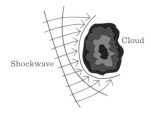

Around 6 billion years ago, as the dwarf galaxy veered into the Milky Way, still regions of hydrogen gas were quickly and violently disrupted. For half a billion years, this impact rippled out across the galaxy. As the shockwaves created by the impact encountered clouds of interstellar gas, they didn't just knock them out of the way, instead they acted more like a shepherd, herding the thinner outer reaches of the clouds inwards by pushing at them from all sides. The result of this was that ancient clouds became compressed, increasing in density, and in some cases they became compacted enough to create the right conditions for a new generation of star formation to begin. Repeated across the galaxy, the impact of Sagittarius Dwarf's encroachment into our galaxy reverberated in the most spectacular of ways – thousands upon thousands of dormant nebulae sparking into life. This was the pillars of creation on a galactic-wide scale, the whole Milky Way shimmering in the light of a new age that would bring with it the most fundamental change of all.

In the midst of all this star formation, one seemingly insignificant star would appear on an outer arm of our galaxy. This yellow dwarf star's appearance seems to have coincided directly with the period of star formation triggered by the collision of Sagittarius Dwarf. Now we can't say for certain that the collision with Sagittarius Dwarf caused the formation of our sun; the data is not precise enough and our understanding is not deep enough for that. But what we can say is that the birth of the Sun coincided with enhanced rates of star formation in the Milky Way caused by that collision. So, it certainly is a possibility that the formation of the Sun, of the Earth, of life, and ultimately our civilisation, can be traced back to the most violent of moments in our galaxy's history.

The poet John Donne famously wrote, 'No man is an island, entire of itself. Every man is a piece of the continent, a part of the main.' By which he meant that no human being can isolate themselves from the rest of humanity, because our origins and our fates are so deeply intertwined, and therefore we must care deeply for each other. The same is true for galaxies. No galaxy is an island entire of itself. What's more, the history of the Milky Way stretches back 13 billion years. That's pretty much the entire history of the Universe. Its story is one of collisions and interactions between galaxies, of rivers and flows and streams of stars, stirring up the void and triggering the formation of worlds like ours. You, me, everyone, can trace our origins back to a collision between galaxies. You may be small, but you are a consequence of grand events.

And those grand events haven't ceased; it just feels like it, because we don't perceive events that play out over billions of years involving billions of stars. But the unique thing about this time in our history is that we can speak with some confidence not only about our galaxy's past, but also about our galaxy's future. And just as inexorably as the great islands of stars drift across the Universe, change will come again.

Left: The Red Spider Planetary Nebula gets its shape and colour from the interaction of stellar wind and supersonic shockwaves.

A SPIRAL STRUCTURE

For all of the detail that Gaia is giving us in mapping the Milky Way, it is still difficult to imagine the pure beauty and majesty of the galaxy we call home. We are stuck in the middle of this island of hundreds of billions of stars, and even with our most advanced technology our view remains limited, confined to seeing the immense structure from within. To get a more complete perspective of our galaxy we need to gaze outwards to other islands to catch a glimpse of ourselves reflected within them. To objects like the Pinwheel galaxy (Messier 101), a spiral galaxy 21 million light years from Earth. Or UGC 12158, a barred spiral galaxy 384 million light years from Earth that is often referenced as a similar galaxy in structure to our own. We can't be certain that any of these is our galactic twin, but we can look at objects like this and learn a lot about the structure of our own galaxy.

NGC 3949, like our Milky Way, is a galaxy that has a bright central bulge consisting of older stars surrounded by a disc of younger blue stars and nebulae. If we could peer back at the Milky Way from 50 million light years away, perhaps this is what we would

ORBITING SAGITTARIUS A*
As Einstein predicted, the elliptical orbit of a star like S2 around a black hole does not remain stationary, but advances, creating a rosette shape.

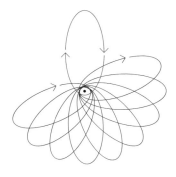

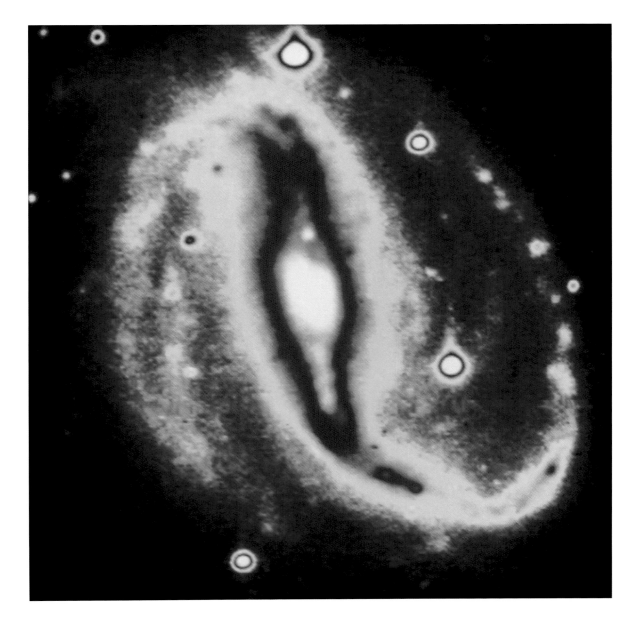

Right: The barred spiral galaxy NGC 7479. Red and white indicate old stars, while yellow represents younger stars and clouds of ionised hydrogen.

Below: The Hubble Sequence for the classification of galaxies, published by Edwin Hubble in 1936.

see – the most typical type of spiral galaxy, a barred galaxy, with a central bar-shaped structure that acts as a stellar nursery, drawing in gas from the surrounding spiral arms to feed the formation of new stars in the centre. We believe that at least two-thirds of spiral galaxies contain a bar structure like this, but it's a structure that doesn't last; present in adulthood as a galaxy like ours ages and star formation slows, it will gradually disappear as we evolve into a standard spiral galaxy.

Buried within the glare of the bar and at the centre of most spiral galaxies is an object that dominates the dynamic of the entire Milky Way. Sagittarius A* is a supermassive black hole at the centre of our galaxy, with a mass at least 4 million times that of the Sun. We can trace its influence by plotting the orbits of a class of stars known as S stars. These orbit closely to Sagittarius, revealing the gravitational power of the monster at the centre of our island. We are pretty certain that supermassive black holes exist at the centre of almost all spiral and elliptical galaxies, and we'll return to these extraordinary structures in much more detail later.

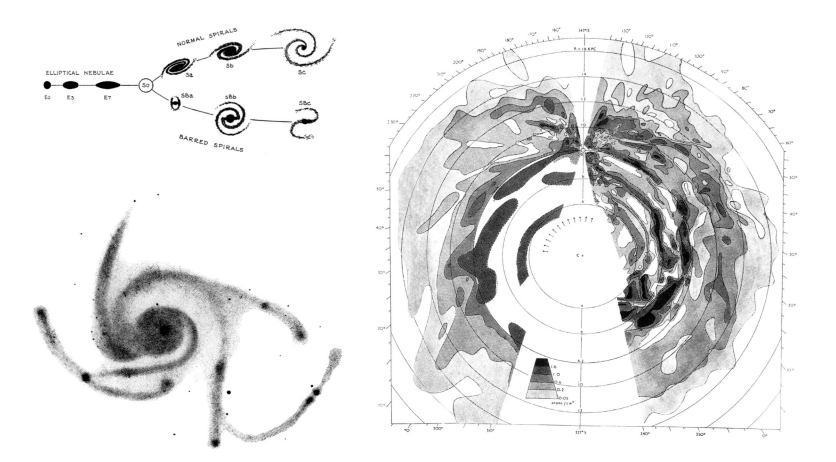

Above: The Pinwheel Galaxy (M101) was thought to be a nebula when it was drawn by 19th-century Irish astronomer William Parsons.

Above: The first diagram to show the distribution of neutral hydrogen in the Milky Way, 1958. Darker shading means denser hydrogen.

Left: Pink dots are hot, dust-obscured galaxies discovered by WISE. Each puts out over 1,000 times more energy than our own galaxy.

'Spiral arms are like traffic jams, in that the gas and stars crowd together and move more slowly in the arms. As material passes through the dense spiral arms, it is compressed and this triggers more star formation.'
Denilso Camargo, Federal University of Rio Grande do Sul, Brazil

So the structure of our galaxy – and perhaps all spiral galaxies – begins with a black hole at the very centre, then beyond this is a central bulge containing the oldest of our galaxy's stars crowded into the galactic centre. (There is still debate around the size and structure of the bulge in our galaxy.) Then we come to the elongated bar-shaped structure, where gas and dust are pulled in and new stars form in vast stellar nurseries. Beyond this lies the most distinctive feature of our galaxy and the place we call home.

Stretching out from the gravitational dominance of the galactic centre are four spiral arms (along which the vast majority of star birth occurs). We still don't know the exact organisation of these grand structures, but we think there are four spiral arms that wind out from the galactic centre – Perseus, Scutum-Centaurus, Sagittarius and the outer arm. From our vantage point here on Earth, it's difficult to peer past the endless clouds of dust that block and distort our view of these arms, leaving vast holes in our understanding of their structure. But recent observations using data from NASA's Wide-field Infrared Survey Explorer (WISE) has enabled us to peer past the dust and see the outline of the spiral arms traced out in the pattern of hundreds of dust-shrouded clusters of stars.

In our galaxy the spiral arms are where most stars are born, expansive structures filled with the gas and dust that is needed for new star formation. This means that the arms are studded with star clusters, stellar nurseries that are embedded in the spiral structures, making them immensely helpful in plotting the shape of the galaxy's spiral arms. (However, these young stars don't stay here long, they live out their youth here before migrating further out into the galactic disc.) Many of these are impossible to see using optical telescopes, due to the dust that shrouds them, but infrared astronomy like NASA's WISE mission has allowed us to plot the position of hundreds of these clusters.

WISE alone has discovered more than 400 of these dust-shrouded nursery stars embedded in the arms of the Milky Way, and it seems to be revealing significant differences between the four arms. Perseus and Scutum-Centaurus seem to be the most populated of the four arms, and perhaps the only two major stellar arms in our galaxy. Brimming with stars both young and old, they outshine the more subdued Sagittarius and outer arms where fewer stars are forming, despite the presence of gas and dust. Why this difference exists we do not know, and there is still much conjecture about the true structure of the spiral arms of the galaxy, how many of them there are and how they entwine. What we do know is that our solar system can be found on a minor spiral arm known as the Orion Arm that sits between Perseus and Sagittarius. We find ourselves on the inner rim of this spur, about halfway along and roughly 26,000 light years from the galactic centre. It's from here that we attempt to draw our view of the Milky Way and the Universe beyond.

Left: Artist's impression of WISE in orbit around Earth.

THE STAR THAT CHANGED THE UNIVERSE

One hundred years ago the Universe was a much smaller place. In the early twentieth century, many astronomers were still convinced the Milky Way marked the boundary of the observable cosmos, with nothing lying beyond this single 'island universe'. One solar system, in one galaxy, in one universe. Arguments among astronomers about the size of the Universe and what lay beyond the boundaries of the Milky Way had rumbled on for decades, and at the centre of the debate was the nature of a particular type of object, a faint patch of light known as a spiral nebula. These diffuse clouds of dust had long been observed by astronomers going all the way back to Messier's cataloguing of the night sky while he was on the hunt for comets. Objects like M51 (the Whirlpool galaxy), when viewed through the most powerful telescopes of the time, appeared to reveal a tantalising spiral shape, a now-familiar structure lurking within the clouds. At this time, the nature of these spiral clouds was far from understood and many believed them to be just one type of gas cloud lurking within the boundaries of the Milky Way, but some astronomers believed they signalled a universe beyond the horizon of our own galaxy. They argued these were not gas clouds within our island but whole other islands of stars, so far away that it was impossible to resolve a single star. It raised the intriguing possibility that these spiral nebulae were islands of billions of stars, distant galaxies waiting to be discovered.

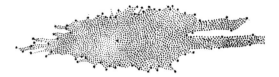

Today, it's difficult for us to look at images like the 1899 photograph of M31, or the Andromeda Nebula, as it was called at the time, and not immediately see the structure of a galaxy, but at the turn of the twentieth century, proving them to be such was far from straightforward. Our technological power would have to evolve before we would be able to resolve any greater detail lurking within these mesmerising structures.

The Andromeda Nebula was of particular interest to astronomers due to its brightness in the night sky, and by the early 1920s evidence had been building to suggest this particular diffuse cloud contained far more structure than a simple nebula. Observations of novae (bright stars that suddenly appear and then slowly fade) within M31 by American astronomer Heber Curtis, revealed that they were significantly fainter than those that occurred elsewhere in the night sky, suggesting they were farther away. He also proposed that the presence of dark lanes that could be observed within the nebulae were structures remarkably similar to the dust clouds in our own galaxy. Combined with the work of other astronomers, Curtis became the lead advocate of the 'island universe' hypothesis, the theory that spiral nebulae like M31 and M51 were in fact independent galaxies beyond our own. The climax of this bad-tempered scientific dispute was a clash between Curtis and his main opponent Harlow Shapley in an event that became known as The Great Debate.

The Great Debate took place on 26 April 1920 at the Smithsonian Museum of Natural History in Washington, DC, with both scientists presenting their case for the shape and scale of the Milky Way and the Universe beyond. Shapley argued that spiral nebulae such as Andromeda simply couldn't be outside of the Milky Way because, according to his (incorrect) calculations, it would mean Andromeda was over a billion light years away, a distance that was unthinkable at the time. Curtis, on the other hand, presented evidence revealing that there were more novae in Andromeda than in the whole of the Milky Way, suggesting M31 was a separate galaxy with its own characteristic frequency and quality of novae events. Despite the compelling evidence presented on either side, The Great Debate would not lead to any significant resolution and the scientific community would remain divided between those trapped inside our galaxy and those who had already stepped outside of it.

It would take the discovery of Hubble Variable number 1 (V1 for short), the most important star in the history of cosmology, to finally resolve the great debate and fling us out of the confines of the Milky Way forever. V1 is a Cepheid variable, a type of star that varies in its brightness in a helpfully predictable pattern. In 1908, American astronomer Henrietta Leavitt had demonstrated that this type of star could be a reliable marker in measuring cosmic distances. By plotting the light curve of one of these stars, the intrinsic brightness can be calculated and therefore its actual distance measured. Known as Leavitt's Law, this provided astronomers with the first 'standard candle' to measure the scale of the Universe, but it would be another 15 years before this technique and the discovery of V1 would allow us to break out of the confines of the Milky Way.

Opposite: The Andromeda Galaxy, photographed in 1899 by Isaac Roberts and labelled 'The Great Andromeda Nebula, M31'.

Top: The shape of the Milky Way based on star counts by William Herschel in 1785; the Solar System was assumed to be near the centre.

Middle: The 'tantalising spiral' of the then-called the Whirlpool Nebula, M51, drawn by William Parsons in 1850.

Bottom: The programme for 'The Great Debate' which was released to the press on 26 April 1920. The results changed astronomy forever.

THE SCALE OF THE GALAXY

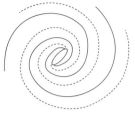
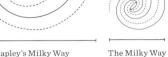

Shapley's Milky Way
300,000 ly

The Milky Way
105,700 ly

Curtis's Milky Way
30,00 ly

On the night of 5 October 1923, at the Mount Wilson Observatory in California, American astronomer Edwin Hubble pointed the most powerful telescope of the time up towards the Andromeda Nebula. Throughout the night and into the early hours of the next morning, the 100-inch Hooker Telescope gazed out into the distant cosmos and captured a series of photographic plates that at first seemed to reveal a trio of potential novae shining within the spiral structure. But when Hubble compared the images with previous exposures of the same stars, he noticed that one of them brightened and dimmed over a relatively short time period – 31.4 days from the beginning to the end of its cycle.

In one of the greatest moments in the history of human exploration, Hubble had stumbled across a Cepheid variable located within the Andromeda Nebula; Hubble Variable number 1 was a star that could allow us to measure the distance to this most controversial of cosmic structures. Running through the calculations, Hubble concluded that this star was around 1 million light years from Earth, a number three times greater than the accepted diameter of the Milky Way at that time. In an instant, Hubble had blown apart our place in the Universe; Andromeda was not a nebula, it was a galaxy, as were the hundreds of other spiral nebulae that had also been observed. The Milky Way was just one galaxy among an untold number stretching out across the Universe. 'Here is the letter that destroyed my universe,' were the words apparently muttered by Shapley when he received a note from Hubble detailing the discovery of V1.

PLOTTING A CEPHEID VARIABLE
Peaks in the plot of a Cepheid variable correspond with the star looking bigger and brighter in the sky.

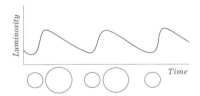

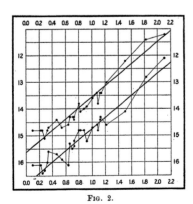

FIG. 2.

Above: Henrietta Leavitt's plot of a Cepheid variable, 1912. The x-axis is the logarithm of time, the y-axis of luminosity.

Right: The Hooker Telescope in Mount Wilson Observatory. Photographed in 1925.

Almost 90 years later, once again the most powerful telescope of the age was pointing its mirrors towards V1. This time, though, the telescope wasn't the Earth-bound Hooker, instead it was Hubble, observing from low Earth orbit, travelling at 7.5 kilometres per second around our planet. Named after Edwin Hubble, the Hubble Space Telescope has provided endless insights and images of the cosmos over its 30 years of service. Staring deep into the Andromeda Galaxy, the Hubble Telescope revealed in unprecedented detail the cycle of V1 from its dimmest to its brightest, a star we now know sits even further away than Hubble first calculated, at around 2.5 million light years from Earth. (It was discovered in 1953 that there are in fact two types of Cepheid variables, with one being dimmer, which resulted in the distance estimates doubling for some of our observations, including V1.) Hubble Variable number 1 is just one of perhaps as many as a trillion stars that make up the stellar population of our nearest major galaxy. Just like the Milky Way, Andromeda, or M51, is a barred spiral galaxy, and at 220,000 light years across – roughly double the size of our own galaxy – it is by far the largest in scale of all the galaxies in the Local Group.

Like distant siblings, the life story of Andromeda very much echoes the story of our own galaxy. Formed from the collision of a cluster of smaller protogalaxies, Andromeda began life as an incredibly bright, highly active galaxy, bursting with the light of new star formation. Just like the Milky Way, its shape and character has been heavily influenced by a series of collisions and near-misses with satellite galaxies across its lifetime. Today, its structure is littered with the evidence of these interactions: encircling the galaxy are a number of giant stellar streams, a halo of stars (similar to our own Sagittarius stream) that are the remnants of dwarf galaxies or globular clusters that were ripped apart by the might of Andromeda. We also see the evidence of other more recent collisions, perhaps as little as 100 million years ago, in the counter-rotating disc of gas and young stars that we have observed in the centre of Andromeda, a feature that is reminiscent of the trail of stars left behind in our galaxy by the Gaia-Enceladus collision.

But perhaps the most dramatic event in Andromeda's 10-billion-year history was when the Triangulum Galaxy (M33), the third-largest galaxy within the Local Group, grazed past Andromeda around four billion years ago and, just like Sagittarius Dwarf's impact on the Milky Way, drove a shockwave through the galaxy that led to a period of intense and sustained star formation. We cannot begin to know what impact the formation of billions of new stars had on the formation of the billions of new planets that we know must have come into existence during this period in Andromeda's history, but it's fascinating to think that a similar collision in our own galaxy led to the formation of a planet that would go on to nurture not just life but a civilisation.

Next time you are lucky enough to stand under the clearest of moonless night skies, look up (with the help of the multitude of online star guides) and try to find the fuzzy patch of light that is Andromeda. One of the most distant objects that can be seen with the naked eye, you'll be staring across the Universe, towards a galaxy of a trillion stars, trillions of planets and perhaps even towards an advanced civilisation that is looking up and back at our galaxy with the same sense of wonder.

TRIANGULUM

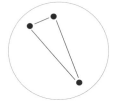

Top: The Cepheid variable RS Puppis, swaddled in a cocoon of reflective dust. RS Puppis is 10 times more massive than our sun.

Above: Triangulum (lower left) and Andromeda (upper right) Galaxies, separated by the red giant star Mirach (Beta Andromedae, centre).

A NEIGHBOUR COMES KNOCKING

Set apart at a distance of 2.5 million light years, it's impossible to imagine humans ever travelling to the distant worlds of Andromeda. Our own galaxy is so vast that exploring this island may never be possible even if we put together a fleet of robotic explorers and send them out into the darkness. But where we cannot go, we can often see, and once again the Hubble Space Telescope has enabled us to stare way beyond our own cosmic shore, looking deep inside our neighbouring galaxy to reveal its extraordinary breadth and beauty.

In 2015, Hubble took the largest image it has ever assembled in its 30-year exploration of the cosmos. A view of the image is printed here, but to truly appreciate the scale of this picture, follow the link esahubble.org/images/heic1502a/zoomable, where you will be able to explore the 100 million stars captured in this 61,000-light-year-long stretch of Andromeda. A composite image compiled from 7,398 separate exposures, this reveals the structure of the galaxy from the galactic bulge in the centre to the spiral arms stretching through an expanse of swirling dust that punctures the swathes of stars, both young and old. With the help of Hubble, we are beginning

'By getting images of beautiful
coloured planetary nebulae from
Hubble, we have been able to see
that our universe is a beautiful
painting of colours.'
*Rana Ezzeddine, Astronomer,
University of Florida*

Below: A crop of Hubble's
Andromeda image has 1.5
billion pixels. It would take
more than 600 HD TV screens
to display the whole image.

to map not just our island of stars but our neighbouring island, too, and in doing so Hubble is allowing us to see beyond the billions of stars to observe a massive structure within the Andromeda Galaxy that is entirely invisible from Earth. It's a structure that has enabled us to glimpse a grand dance that has begun between Andromeda and the Milky Way, one that will define the future of both galaxies.

The galactic halo is a vast and ethereal structure that surrounds almost every galaxy, including our own. In a spiral galaxy (like the Milky Way or Andromeda) the halo forms a colossal sphere around a thin galactic disc. Made up of stars and globular clusters towards the centre and great swathes of gas, dust and dark matter as it extends outwards, directly exploring these ghostly structures is almost impossible to our eye – even through a powerful telescope the halo is invisible. Instead, scientists have to use an indirect technique that utilises the effect the halo has on light travelling through it from distant objects like quasars. It's this technique, conducted on an unprecedented scale over the last couple of years, that has allowed us to explore Andromeda's giant halo in extraordinary detail.

In May 2009, during Servicing Mission 4 of the Hubble Space Telescope, a new instrument, the Cosmic Origins Spectrograph (COS), was installed on the orbiting telescope. Designed to detect ultraviolet light that cannot be seen by the usual ground-based telescopes (due to the effect of the Earth's atmosphere on this wavelength of light), the COS has enabled us to peer at the light from 43 quasars that sit directly behind Andromeda's halo. Quasars are the very distant, very bright cores of galaxies that are powered by massive black holes at their centre and are ideal objects to look at for the absorption patterns of a galactic halo. Using the COS to measure how the light from these quasars is absorbed as it travels through the ionised gases of Andromeda's halo, has enabled a team of NASA scientists in 2020 to measure the composition, size and mass of the halo and reveal the full extent of its structure for the first time.

Andromeda's halo is a dynamic structure that contains many clues to the past, present and future of the galaxy. Formed from two distinct layers, there seems to be an inner shell that is more complex and dynamic than the smoother and hotter outer shell. Huge reserves of hydrogen and helium gas, the fuel for future star formation, exist throughout both these structures, but the inner shell has an abundance of heavier elements such as carbon, silicon and oxygen that we think are the product of supernova activity in the galaxy's inner disc. We believe that these gases are feeding the surrounding halo as they explode and die. But it's the scale of the halo that has really taken us by surprise. The inner shell extends for more than half a million light years around the galaxy, and from the very latest observations it seems that the outer shell extends at least another million light years. It means that the entire halo extends 1.3 million light years from the galaxy and as far as 2 million light years in some directions. To get some sense of the scale of this, if it could be seen with the naked eye from Earth it would fill the sky with its translucent glow.

The realisation that Andromeda's halo is a structure of such scale has also made us re-evaluate the halo that we know must surround our own galaxy. Although we live inside it, we cannot see it or even measure it in the way we have with our neighbouring galaxy, but we know that Andromeda and the Milky Way are remarkably similar in structure and that means that the Milky Way's halo must be of a similar scale. Two and a half million light years apart, these two galaxies, with their two expansive halos stretching out across millions of light years of space, are not distant neighbours; they are touching, in contact with each other and getting closer by the day.

THE LAST DANCE

The life story of our galaxy has been shaped by collisions. Stellar clusters and galaxies that stray too close have repeatedly felt the might of the Milky Way and its ability to destroy and consume vast collections of stars. Across its 12.5-billion-year history, each of these interactions has been a one-sided fight – each time the Milky Way has been the bigger entity, the dominant force capable of exerting its power over any invader. But that is all about to change.

For decades, we've known that the largest galaxy in our vicinity, Andromeda, is moving in the direction of the Milky Way at speeds approaching 110 kilometres per second. Racing towards us on a galactic collision course, we have struggled to be certain of its exact trajectory, unsure if we faced a head-on collision or a mere glancing blow. But over the last ten years, with the help of the Hubble Telescope and more recently the Gaia Telescope, we have been able to build up an increasingly accurate picture of the motion of all the galaxies in the Local Group, and particularly the destiny of Andromeda and the Milky Way. Although primarily designed to observe the stars of our own galaxy, Gaia has been able to measure the position and motion of hundreds of stars in our neighbouring galaxy with an unprecedented level of accuracy, allowing us to plot its course both now and stretching billions of years into the future.

Around 4.6 billion years from now, the Sun will have spent all of its fuel, existing only as a dying star, a red giant, expanding out far into its realm and engulfing Mercury, Venus and possibly even the Earth. By this time, the Solar System would have reached the end of its days, but while it withers, a much bigger game will be at play as the Milky Way finally meets its match. Perhaps our planet will cling on just long enough, a tortured lifeless shell, to witness this final act of the galaxy that we have called home. Colliding not head on but with a hefty glancing blow, the impending arrival of Andromeda will result at first in a colossal burst of activity as the vast reserves of gas within the Milky Way are stirred into a new era of star birth. But this final age of fertility will not last, these new stars will mark the beginning of the end for the Milky Way as the collision with Andromeda slowly pulls apart the ancient spiral structures just as our galaxy has devoured smaller galaxies in the past.

The Milky Way won't die in the collision, it won't be destroyed, but it will lose its identity. With great majesty and great violence, the two galaxies will slowly pull each other apart, scattering stars until no trace of the grand spiral structures remain and they become merged into a single, larger entity. Despite the scale of the collision, we expect almost every star to survive. Even with a full-on impact, the chances of even

A final almighty collision

Andromeda and the Milky Way are approximately the same size. When they come together, sparks fly. You end up with something completely different. You have a huge burst of stars. Most of the gas in the galaxy gets heated up and destroyed in some sense. And you end up with something that looks fundamentally different to either Andromeda or the Milky Way.

We think that the collision between Andromeda and the Milky Way will be the most substantial collision in the history of both those galaxies, unlike the Gaia-Enceladus collision, which did change the Milky Way a little bit, but involved a much smaller galaxy compared to our own galaxy. The end result will be something that is quite different. After the collision with the Andromeda Galaxy, the Milky Way, as we know it now, will not exist anymore.
David DeSario, Astronomer, Durham University

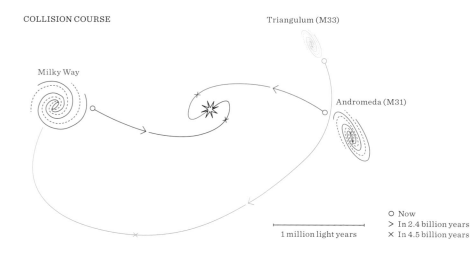

COLLISION COURSE

Triangulum (M33)

Milky Way

Andromeda (M31)

○ Now
> In 2.4 billion years
✕ In 4.5 billion years

|—————| 1 million light years

Below: This series of illustrations shows what a collision between Andromeda and the Milky Way Galaxies could look like from about 25,000 light years away from the centre of the Milky Way. The future view from our solar system will most likely be very different, depending on how the Sun's orbit within the galaxy changes during the collision.

2 billion years: The disc of the approaching Andromeda Galaxy is noticeably larger.

3.75 billion years: Andromeda fills the field of view. The Milky Way begins to show distortion due to tidal pull from Andromeda.

3.85–3.9 billion years: During the first close approach, the sky is ablaze with new star formation, which is evident in a plethora of emission nebulae and open young star clusters.

4 billion years: After its first close pass, Andromeda is tidally stretched out. The Milky Way, too, becomes warped.

5.1 billion years: During the second close passage, the cores of the Milky Way and Andromeda appear as a pair of bright lobes. Star-forming nebulae are much less prominent because the interstellar gas and dust has been significantly decreased by previous bursts of star formation.

7 billion years: The merged galaxies form a huge elliptical galaxy, its bright core dominating the nighttime sky. Scoured of dust and gas, the newly merged elliptical galaxy no longer makes stars and nebulae appear in the sky. The ageing starry population is no longer concentrated along a plane, but instead fills an ellipsoidal volume.

two stars colliding is minuscule; the distances between stars means that galaxies are much too diffuse for that. But many stars will have their orbits disrupted, and some stars, including our sun, will have their orbits disrupted so much that they will suffer a strange and poignant fate. Models suggest that the force of the collision could fling the remains of the Solar System to the outer reaches of the new galaxy, or even entirely eject it from the new galaxy all together. Even if this is not the fate of our sun and whatever planets remain around it by then, we do know this will be the fate of some stars. Sent to the fringes of this new merged galaxy, the unlucky few will be ejected into the emptiness of space and banished forever, leaving behind a new vast elliptical galaxy of perhaps a trillion or more stars.

No galaxy, not even this grand new structure, can remain an island unto itself for ever. Around 150 billion years from now, the remaining galaxies of the Local Group are predicted to merge into a single vast galactic entity, and perhaps only then will our island be alone.

Our universe is expanding, every galaxy rushing away from each other, and in the future they'll be rushing away even faster, until eventually there will come a time when, if we sent a beam of light out to any galaxy, it would never catch it and so consequently the light from any galaxy outside our own would never arrive. The astronomers of the far future might imagine that they live in a universe populated by countless billions of galaxies, but they won't be able to prove it – they won't be able to see a single one. We are the lucky ones, living in an age where we are able to see the true scale and majesty of a universe filled with endless islands of stars.

2 BILLION YEARS   3.75 BILLION YEARS   3.85–3.9 BILLION YEARS

4 BILLION YEARS   5.1 BILLION YEARS   7 BILLION YEARS

BLACK HOLES

'Deep into that darkness peering,
long I stood there, wondering, fearing,
doubting, dreaming dreams no mortal
ever dared to dream before.'
Edgar Allan Poe

MAELSTROM IN THE MILKY WAY

Black holes – elusive, destructive monsters steeped in mystery. These are bodies that challenge our notions of space, time and the Universe, and teach us that the laws of physics that we once regarded as immutable are anything but. However, black holes aren't just weird aberrations of space for physicists to argue over; we now think these objects had a pivotal role in the history of the Universe. Their reputation as the ultimate cosmic monsters, destroying anything that strays too close to them, isn't unearned, but that's only half the story. Black holes are also agents of creation: they sculpt galaxies, they bring stars to life, and one might even have helped forge our solar system, enriching our corner of the Milky Way with the right ingredients to create this beautiful planet that we call home.

Look up into the night sky and we can see the planets that populate our solar system, the stars that fill the Milky Way, and we now know that hiding around those stars are billions of alien worlds waiting to be explored. But for all of the scale and grandeur of our galaxy there is much that is still hidden, dark and unseen. Turn your gaze towards the constellation of Sagittarius and you are peering into

Opposite: Dense gas and dust surrounds the centre of the Milky Way (left), orbiting the supermassive black hole, Sagittarius A*.

What is a black hole?

Black holes, as far as we know, are remarkably quiet, but don't let that fool you, because what's going on immediately outside of them is rip-roaring activity. So if one were ever unfortunate enough to fall into a black hole, you would never get back out, and you would find it is probably a quiet, boring little place – until you were ripped to shreds. But it's revving up matter to such extraordinary energies that all these crazy, violent things are happening just outside it; a cosmic light show that is amongst the most violent interactions we've been able to see in the cosmos.

David Kaiser, Physicist, MIT

the heart of the Milky Way, a view that is blocked and disrupted by vast swirling clouds of dust and gas. To peer into this realm, we need the help of our most powerful explorers, the telescopes that have left the blurring blanket of the Earth's atmosphere behind and that look from their own orbits deep into the cosmos.

When the Hubble Space Telescope opened its infrared eye towards Sagittarius it was able to look through the clouds of dust and return an image of unimaginable scale and beauty. Here we see stars packed together, a seething maelstrom a hundred times denser than our own solar neighbourhood. And inside the Milky Way's inner sanctum, 25,000 light years from Earth, these stars perform a bizarre frenetic dance. Spinning around at 7,500 miles a second, 400 times the speed at which the Earth orbits the Sun, these stars seem possessed by a hidden force. For all of the stellar beauty and structure that this image reveals, the most famous object contained within it is the one we cannot see, a supermassive black hole at the heart of the Milky Way, which we call Sagittarius A*.

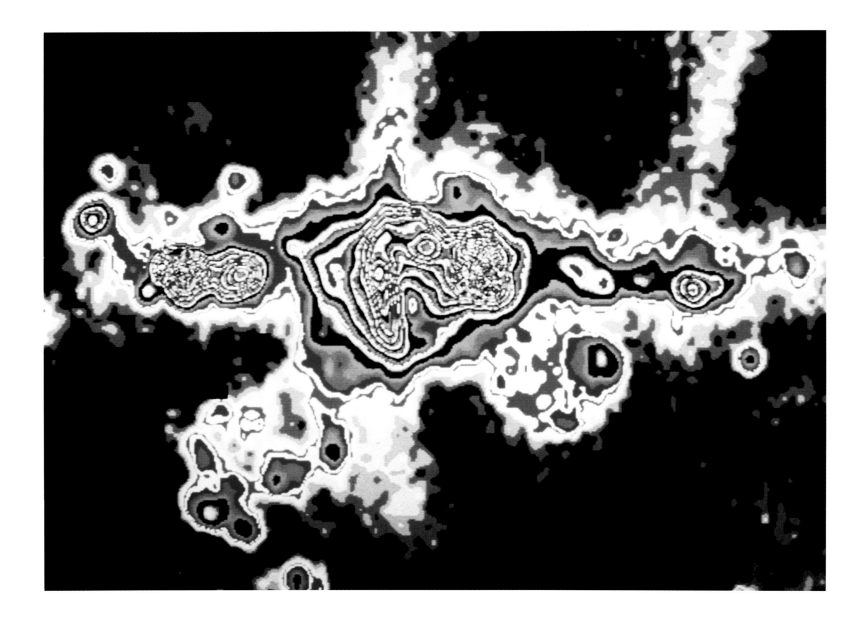

Before   After

THE MONSTER STIRS

On 14 September 2013, the Chandra X-ray Observatory turned its gaze towards the heart of the Milky Way. Launched on 23 July 1999 aboard the Space Shuttle Columbia, Chandra is one of NASA's four great observatories (along with Hubble, Spitzer and the now deorbited Compton Gamma Ray Observatory, or CGRO) and is designed to allow scientists to see a universe that is invisible to human eyes.

Chandra is the most sensitive X-ray telescope ever built, circling the Earth 200 times higher than Hubble, over one-third of the way to the Moon, in a highly elliptical 65-hour orbit. It is this trajectory that allows the telescope to continually observe the Universe for around 55 hours of each revolution. Here on Earth, even our most powerful telescopes are unable to conduct X-ray astronomy due to the absorption of these powerful light rays by the planet's upper atmosphere. Chandra is able to observe X-rays from clouds of gas that are so vast it takes 5 million years for light to travel from one side to the other, and gives us a precious, unprecedented view of the Universe, allowing us to explore the characteristics of many star types as well as cosmic phenomena such as galaxy clusters, neutron stars, supernova remnants and, of course, black holes. Designed to detect the high-energy X-rays emitted by the hottest, most violent regions in the Universe – created when matter is heated to millions of degrees – Chandra is our first black hole explorer.

Focused towards the centre of the galaxy in 2013, the telescope was monitoring a cloud of interstellar gas known as G2 and the X-ray bursts it produced, but during these routine observations Chandra chanced upon something on a completely different scale. Chandra had caught sight of an extraordinarily huge flash of X-rays coming from the galactic core. The monster at the heart of the Milky Way had awoken, and we were in the right place at the right time to capture the moment.

Chandra had observed the largest X-ray flare ever detected from near Sagittarius A*. For just a brief couple of hours, something was going on that produced a flash of X-rays 400 times brighter than in its usual quiet state. We are still not exactly certain what caused this eruption, but the most likely explanation is that we captured the black hole feeding.

If we could travel beyond the centre of our galaxy, tightly packed stars spinning around the invisible abyss would give way to a vast frozen field of asteroids. With trillions of icy worlds orbiting at immense speeds, this is a no-man's land where every rock is in a dance with a monster, flirting with the colossal forces that grasp anything that passes in close proximity. And, of course, when one of these tortured worlds strays too close, the outcome is inevitable. Torn apart by the colossal gravitational forces, the rocky debris would reach immense temperatures as it edged towards the point of no return, producing the explosive flash of X-rays that Chandra was able to witness, before it disappeared forever across the threshold.

At 44 million kilometres wide, with a mass 4 million times that of the Sun, Sagittarius A* is supermassive, a glowing ring of hot gas and dust spinning at 60 miles a second, where the gravitational pull is so strong that all of the colour and beauty of the cosmos is drained out of sight, so not even light can escape.

The flash that Chandra saw was Sagittarius A* having the smallest of snacks, a rare moment where this brooding monster revealed a glimpse of its power in the darkness. But no scientist can describe exactly what happened to that rock. Despite a century of theoretical and observational exploration, we are only at the beginning of our journey towards a deeper understanding of the black hole.

X-RAY TELESCOPES
X-rays can pass through most mirrors (called intermediate incidence), but if the mirror is at just the right angle (the grazing incidence) the X-ray will be reflected. To achieve this, an X-ray telescope has to turn its mirrors nearly parallel to the incoming light. But this leaves a large hole in the middle of the telescope, so it misses a lot of X-rays. To solve this, X-ray telescopes use cylindrical mirrors and nest one inside the other.

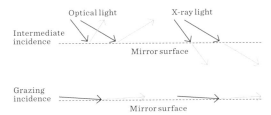

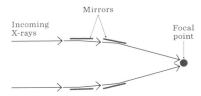

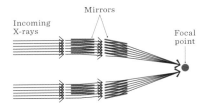

Opposite: The 2013 X-ray flare from Sagittarius A*, 400 times brighter than any previously observed flares.

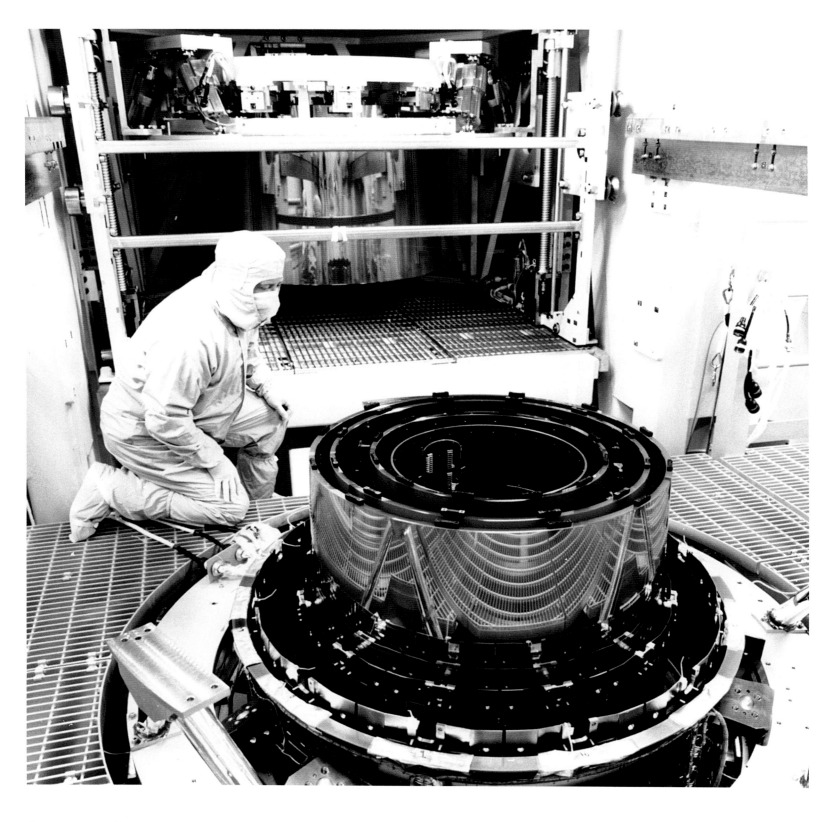

Above: The two sets of four
nested mirrors of the High
Resolution Mirror Assembly
(HRMA) of the Chandra
X-Ray Observatory.

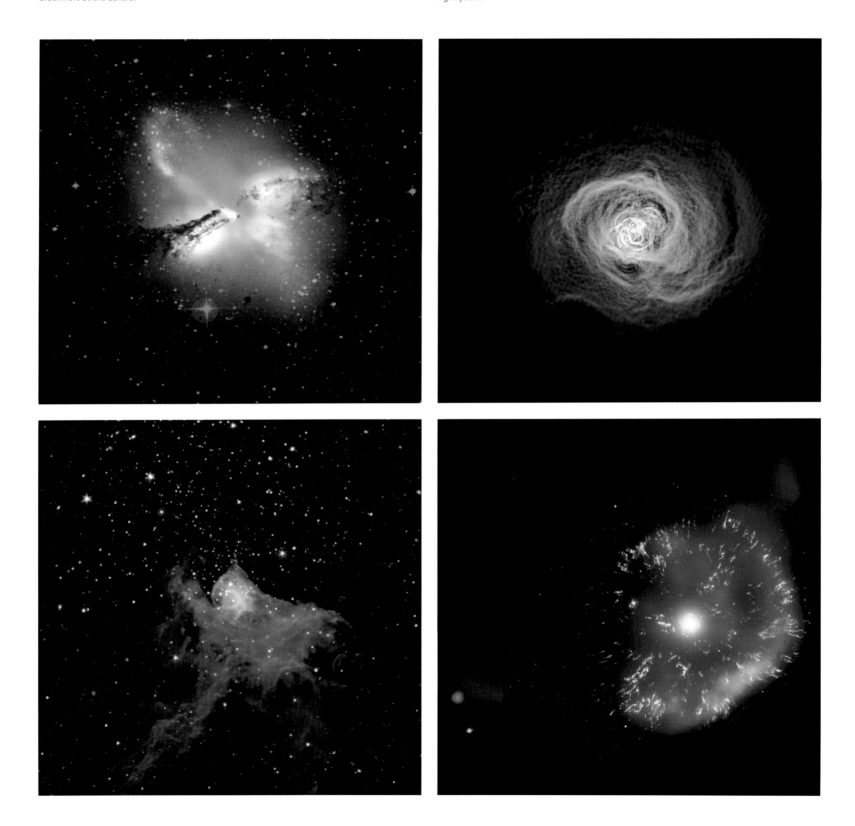

Below: Chandra captures the remains of a 10-million-year-old explosion in Centaurus A, still blasting jets from the black hole at the centre.

Below: A vast wave of hot gas in the nearby Perseus Galaxy cluster, discovered by Chandra, spans 200,000 light years.

Above: The star-forming molecular cloud Cepheus B, in the Milky Way, imaged with combined Chandra and Spitzer data.

Above: Chandra image of GK Persei, which caused a stir in 1901 when it appeared as one of the brightest stars in the sky for a few days.

THE FINAL FATE FOR OUR STARS

'There should be a law of nature to prevent a star from behaving in this absurd way.'
Arthur Eddington

Every history of the black hole begins with the name John Michell, an eighteenth-century clergyman, geologist and natural philosopher. His letter to Henry Cavendish, published in the Philosophical Transactions of the Royal Society in 1784, described the theoretical possibility that a body so massive could exist that even light could not escape its grasp. With a few simplistic calculations, Michell had given birth to the idea of a dark star – a looming, invisible presence so powerful it could trap light within its gravitational grasp, and an object so massive it could only be detected by the influence that its vast gravitational force had on any objects that approached it.

Michell's dark stars intrigued many of the great astronomical minds of the time, but an object without a chance of observation was far less seductive to chase than the myriad phenomena that were appearing in the eyepieces of the ever more powerful telescopes of the time. Lost in the annals of science, Michell's contribution to the history of dark stars would not be rediscovered until the 1970s.

To appreciate the modern foundation of our understanding of dark stars, we need to begin our story at one of the darkest moments of the twentieth century. In December 1915, somewhere on the Eastern Front, a German lieutenant in the artillery corps sat huddled in a damp trench reading the most influential scientific manuscript of that century – and perhaps even of all time. At the age of forty, older than almost all the soldiers around him, Karl Schwarzschild had chosen to join the German war effort rather than observe it unfolding from the more protected position of the Astrophysical Observatory in Potsdam, where he had been serving as director before enlisting.

NY484-12/26-PRINCETON,N.J.:Though the layman's reaction will probably be "It's Greek to me," this is a new "Generalized Theory of Gravitation," developed by physicist Albert Einstein,author of the theory of relativity.The theory attempts to interrelate all known physical phenomena.The new theory may well rank with the original publication of relativity as a milestone of scientific achievement,intended to bring relativity and the quantum theory into a single system. It is impossible to say certainly at this time whether the theory is successful.Preliminary indications are favorable. ACME TELEPHOTO

The Four Equations

The heart of the generalized theory of gravitation is expressed in four equations, shown in the accompanying illustration.

$$g_{ik;l} = 0 \; ; \; \Gamma_{l} = 0 \; ; \; R_{ik} = 0 \; ; \; g^{,\nu}_{,\varsigma} = 0 \qquad \text{German lower case} \; \sigma$$

The equations have the mathematical properties which seem to be required in order to describe the known effects, but they must be tested against observed physical facts before their validity can be absolutely established.

ANATOMY OF A BLACK HOLE
Singularity is the infinitely dense gravitational centre of the black hole. The Schwarzschild Radius defines the distance between it and the event horizon, the boundary beyond which not even light can escape. Beyond, the photon sphere is an area with gravity strong enough that photons are forced to travel in orbits, in theory enabling you to see the back of your own head.

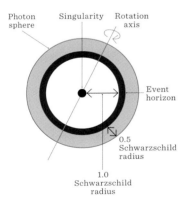

Photon sphere — Singularity — Rotation axis

Event horizon

0.5 Schwarzschild radius

1.0 Schwarzschild radius

In his hands were four landmark papers that had recently been published in the Proceedings of the Prussian Academy of Sciences, which laid out the mathematical foundations of Albert Einstein's Theory of General Relativity. Einstein and Schwarzschild were colleagues and friends, but at this moment, towards the end of 1915, their lives could not have been further apart. In Berlin, Einstein was reaching the pinnacle of his powers, with international fame and fortune just around the corner, while Schwarzschild was entering the last year of his life, scrabbling in the mud of a world war that would leave him with pemphigus, an untreatable skin disease that would ultimately kill him. Despite existing in two completely different universes, for a brief moment these two great minds would become joined by the deepest of questions thrown up by Einstein's paper on 'The Field Equations of Gravitation'. Einstein's field equations describe the relationship between spacetime and matter in the very simplest terms, expressing mathematically the fundamental basis for how gravity functions in the Universe, but at the point of publication Einstein alone was not able to reconcile every mathematical connotation of his Theory of General Relativity.

In the circumstances, it is almost impossible to imagine Schwarzschild sitting in the dim light of a trench, with the carnage of battle all around him, tearing into the mathematical conundrum presented by Einstein's work and coming up with the first known solution to the field equations within the space of a few war-torn days. On Christmas Eve 1915, Einstein received a letter from Schwarzschild detailing his solution. 'As you see, the war treated me kindly enough, in spite of the heavy gunfire, to allow me to get away from it all and take this walk in the land of your ideas,' wrote Schwarzschild at the end of his now-famous letter.

It would be the last flicker of genius that Schwarzschild would share with the world, but its impact would reverberate long into the twentieth century. Schwarzschild's solution to the field equations describes a universe where spacetime is warped by the presence of every massive object, like a planet or a star. But the solution to these equations does not just take us to a new understanding of the clockwork orbits of the heavens, they also predict a far more extreme universe. A universe where the existence of tiny but massive stars can warp spacetime with such force that the gravitational field becomes so powerful that there is a point where every particle – even light – can no longer escape. In the course of his investigations, Schwarzschild concluded there was a radius within which the gravitational attraction between the particles of a body must cause an irreversible gravitational collapse. This phenomenon, known as gravitational radius, or more commonly as the Schwarzschild Radius, is thought to be the final fate of more massive stars in the Universe.

Schwarzschild had created the theoretical possibility of a singularity, a point in the Universe where Einstein's spacetime became infinitely powerful. For the next decade or so, Schwarzschild's work remained a mathematical curiosity, a phenomenon that was simply too strange to exist in the actual observable universe. It would take another academic outsider lost in his own distant universe to bring the dark stars of Schwarzschild's imaginings crashing down to Earth.

Opposite: It was mathematicians who predicted black holes, however unlikely they seemed to astronomers.

Top: Einstein's 1947 Generalised Theory of Gravitation brings relativity and quantum theory into a single system.

In July 1930, on a ship bound from India to England, a young physicist called Subrahmanyan Chandrasekhar sat with his head far away from the oceans that surrounded him on all sides. On his way to take up the scholarship he had won to study physics at Cambridge University, Chandrasekhar used the long journey to travel in his head far away from Earth and explore a fundamental question about the life of a star. Chandrasekhar wanted to understand what happens when a star runs out of fuel, not by looking up at the stars but by playing with the mathematical possibilities created by Einstein and Schwarzschild's work.

At this time, our understanding of the life and death of a star was still very patchy, but the basics of stellar evolution were beginning to take shape. We knew that during a star's life it held an equilibrium between the inward forces of gravity and the outward forces created by the nuclear fusion in its core. But towards the end of its life, this equilibrium was lost and a star could collapse in on itself. Evidence for this had been observed in the night sky with the discovery of tiny dense stars known as white dwarfs – the burnt-out remnants of long-dead stars. These white

Opposite: Subrahmanyan
Chandrasekhar's calculations
directed scientists to uncover
the secrets behind the birth
of the mysterious black hole.

Below: The stellar near-death
experience that formed the
Homunculus Nebula was similar
to a supernova, but stopped
just short of destroying the star.

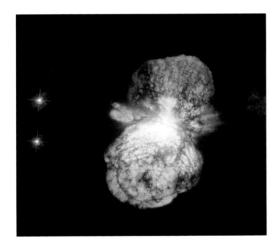

dwarfs were so dense that the very atoms that they were made of were on the edge of being crushed out of existence, an idea that at this time was thought to break the laws of quantum physics and so was dismissed as an impossibility. However, with a brief but beautiful mathematical flourish, Chandrasekhar was about to rock that astronomical boat.

As the ship sailed towards the grand ocean of the Atlantic, Chandrasekhar began to play with the mathematical fate of different-sized stars in the Universe. Using the physical characteristics of a white dwarf as his starting point, Chandrasekhar began calculating what would happen to different-sized stars as they collapsed at the end of their lives. The answer for a star like our sun was straightforward: it would eventually collapse in on itself, forming the strange but familiar structure of a white dwarf, the atoms packed so tightly together it would become 200,000 times denser than the Earth. But when Chandrasekhar started playing with the fate of larger stars, he hit a mathematical anomaly. The calculations suggested that the destiny of stars with a big enough mass would not be a white dwarf but something even more exotic. As bigger stars collapsed, the equations suggested that the overwhelming gravitational forces would cause the atoms to collapse inwards at velocities approaching the speed of light, totally imploding in on themselves, crushing each other to a point of no return, a point where the very matter that had once existed would disappear down its own gravitational well. This, according to Chandrasekhar's calculations, was the potential fate of any star that was at least 1.44 times the mass of our sun, the critical mass beyond which the gravitational forces created by its collapse could not be contained by any atomic structure.

Yet, despite the precision of his calculations, the structure that they predicted was far too outlandish for most in the astronomical community to accept, and therefore the findings were dismissed as a theoretical anomaly. The mathematics of a talented young nobody would not change our understanding of the Universe; the imaginations of even the most respected of astronomers could not stretch to allow such phenomena to exist. 'There should be a law of nature to prevent a star from behaving in this absurd way,' quipped Arthur Eddington in January 1935, as he presented his paper on the theory just after Chandrasekhar had delivered his own results to the Royal Astronomical Society.

By the time Chandrasekhar collected his Nobel Prize in Physics in 1983 (53 years later!) for his studies on 'the physical processes important to the structure and evolution of stars', the calculations were no longer controversial and his conclusion, which became known as the Chandrasekhar Limit, became a cornerstone in our understanding of the fate of stars.

In a flash of theoretical brilliance, Chandrasekhar had used mathematics to paint a picture of a universe where the biggest stars did not just fade into ghosts but imploded into the most powerful entities imaginable. Alone on a ship, with nothing more than a pen and paper, Chandrasekhar had discovered the birth story of a structure that today we call a black hole.

Opposite: Ultraviolet image
of the Helix Nebula, the nearest
example of what happens to a
star like our own sun at the end
of its life.

A BLACK HOLE IS BORN

'We are far from unlocking all the secrets of our galaxies. What's happening on the inside and at the event horizon of Sagittarius is where our mathematical maps seem to end.'
David Rosario, Astronomer, Durham University

The history of Sagittarius A*, the black hole at the centre of our galaxy, begins 13 billion years ago, not long after the birth of the Universe. This was a time when the cosmos was littered with massive stars, burning blue with their intense heat (see Chapter 2), living fast and furious lives, but it was in death that they would express their most powerful form.

One such star, at least 50 times the mass of our sun, burned bright within the embryonic structure that would one day become our home galaxy, the Milky Way. This star rapidly used up the hydrogen and helium that fuelled its frenzied life until after just a few million years there was nothing left to burn, and as the fusion reactions slowed in its core so the delicate balance of the star's structure began to tip. The outward force created by its active core slowly diminished until the crushing inward force of gravity became too much to bear and the star collapsed in on itself at unimaginable speeds before rebounding in a magnificent supernova explosion.

The galaxy would have been filled with the light of this massive blast, a spectacle of unparalleled beauty. As the dying star jettisoned its shell into space at 40,000 kilometres a second, sending out its precious elements to seed the galaxy beyond, all that would have been left behind was its stellar core, a core so massive it would have been unable to resist the overwhelming force of gravity and so would have collapsed in on itself, crushing the atoms within. The remnant of this once-massive star would become tiny but incredibly dense. So dense that it would transform the very fabric of the Universe around it and give birth to a black hole, our black hole – Sagittarius A*.

To understand how the collapsing core of this ancient star created the black hole at the centre of our galaxy, we need to go back again to 1915 and the birth of a new view of the Universe that was created by Albert Einstein and his Theory of General Relativity. Black holes aren't the result of some weird and wonderful transformation of matter, they're a product of the fundamental structure of the cosmos, a structure that Einstein revealed for the very first time. Turning the Newtonian view of an infinite universe on its head, Einstein imagined the fabric of the Universe not as a rigid entity but as flexible and changeable. In Einstein's universe, space and time are inescapably entwined, and it's within this structure, known as spacetime, that the characteristics of a black hole can exist.

In Newton's universe, gravity is an invisible force that acts between any two objects with mass, with the force of gravity defined by the mass of those objects and the distance between them. But Einstein saw things differently. His great insight was to realise that objects with mass are not mysteriously attracted to each other but in fact influence each other by distorting the very fabric of space and time. So the Earth orbits the Sun because it is caught in the distorted spacetime created by the Sun's enormous mass; the Moon orbits the Earth for the same reason and we are pulled downwards towards the centre of the planet because of the distortion of spacetime that is created by the Earth's mass.

Now imagine we take something much more dense, much more massive, than the Earth or the Sun, like the remnant of a huge imploding star. As the stellar core collapses and its density grows exponentially, the distortion of spacetime becomes far greater, producing not so much as a dent in spacetime, but a hole punched right through it. Stray too close to such an object and there's no resisting its intense gravitational pull – not even light, the fastest thing in the Universe, can escape.

Opposite: One of the most violent events in the Universe: a pair of neutron stars colliding, merging and forming a black hole.

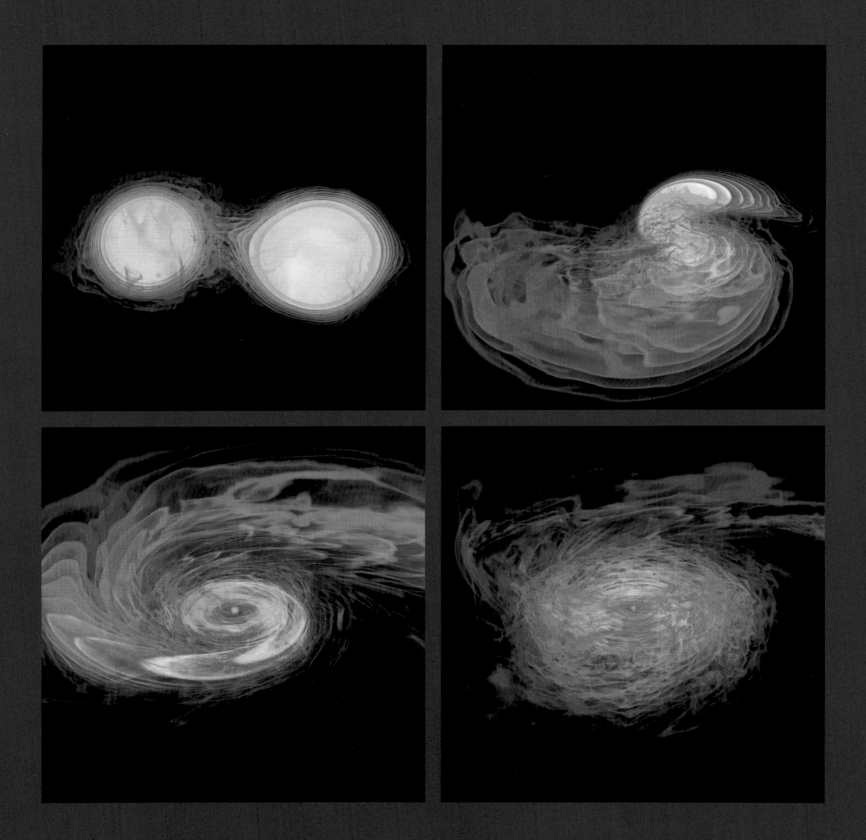

Science News Letter *for January 18, 1964*

ASTRONOMY

"Black Holes" in Space

The heavy densely packed dying stars that speckle space may help determine how matter behaves when enclosed in its own gravitational field—By Ann Ewing

▶ SPACE may be peppered with "black holes."

This was suggested at the American Association for the Advancement of Science meeting in Cleveland by astronomers and physicists who are experts on what are called degenerate stars.

Degenerate stars are not Hollywood types with low morals. They are dying stars, or white dwarfs, and make up about 10% of all stars in the sky.

The faint light they emit comes from the little heat left in their last stages of life. It is not known how a star quietly declines to become a white dwarf.

Degenerate stars are made of densely packed electrons and nuclei, or cores of atoms. They are so dense that a thimbleful of their matter weighs a ton.

Some such stars are predicted in theory to have a density of one million tons per thimbleful. When this happens, the star is essentially made of neutrons and strange particles.

Because a degenerate star is so dense, its gravitational field is very strong. According to Einstein's general theory of relativity, as mass is added to a degenerate star a sudden collapse will take place and the intense gravitational field of the star will close in on itself.

Such a star then forms a "black hole" in the universe.

Modern tools, such as telescopes on an orbiting space platform, may be used to detect such black holes and to help determine how matter behaves when it is enclosed by its own gravitational field.

The light from the most famous white dwarf star, Sirius B, a companion to Sirius—which is the brightest star in the heavens visible from earth—has been captured using the 200-inch telescope atop Mt. Palomar. This was done as part of a program to study at least 20 white dwarfs.

Preliminary analysis of the light from Sirius B indicates that it has an effective temperature of 16,800 degrees Kelvin, or 30,000 degrees Fahrenheit. Its radius can be calculated from the temperature, and is only nine-thousandths that of the sun.

The star must therefore consist mainly of helium or heavier elements.

The speakers at the symposium were Drs. A. G. W. Cameron of the National Aeronautics and Space Administration's Goddard Institute for Space Studies, New York; Charles Misner of the University of Maryland; Volker Weidemann, Physikalisch-Technische Bundesanstalt, Braunschweig, Germany, and J. B. Oke of California Institute of Technology. The symposium was arranged by Dr. Hong-yee Chiu of the Goddard Institute for Space Studies.

• *Science News Letter, 85:39 Jan. 18, 1964*

The black hole conundrum

One of the things that black holes have forced us to reckon with is this very strange, uneasy combination of Einstein's beautiful Theory of General Relativity, which accounts in unbelievable terms for the warping and stretching of space and time, and quantum theory. They don't seem to mesh very clearly at all. In fact, we've been trying for nearly 100 years to kind of knit them together into a single puzzle. That's where this notion of black holes forces us to try harder and harder to see if we can get these two kinds of bedrock notions of modern physics to finally fit together.

It turns out black holes are a remarkable laboratory, at least a theoretical laboratory; they're a mental playground. They force us to grapple with the limits of what we know.

The term black hole was coined in the 1960s, and then the first really compelling evidence was in hand already by the 1970s. And now in some sense, they're kind of banal. Astronomers are kind of finding new ones every day. Now they can hear the ripple through space and time. When two of these collide and merge, we can measure effects of them on the behaviour of individual astronomical systems. They're a feature of the Universe that we now simply take for granted, even though they were considered such an outlandish possibility when they were first thought about.

David Kaiser, Physicist, MIT

So when that enormous star collapsed in on itself 13 billion years ago it punctured the very fabric of the cosmos, its core disappearing into a cosmic sinkhole, and Sagittarius A* was born. Like all holes, black holes have an edge, a boundary that we call the event horizon. As discussed earlier, this is the point of no return; beyond this nothing can escape – not even light – and so the blackness is impenetrable and we cannot glimpse what lies within. But that's just the beginning of a black hole's strange and elusive identity, because not only does the vast mass of a black hole distort space, it also distorts the other part of the fabric of the Universe – time.

Einstein revealed a universe where time does not tick at the same universal rate, instead it changes depending on how the spacetime is distorted around you. The bigger the mass of an object, the greater the impact it has on the passage of time; the closer you are to the centre of any mass, the more slowly time ticks. So here on Earth, time ticks slower standing on the ground than it does in an aircraft flying

Top: The term 'black hole' was first used in print by *Life and Science News* magazine in 1963, and in this article from January 1964.

Right: A black hole strips material from a star, forming a rotating accretion disc that heats up and produces electromagnetic radiation.

THE PULL OF GRAVITY
A gravity well is a model of how spacetime curves around a black
hole. Spacetime is represented as a 2D surface, but in fact it is 3D
and curves around a sphere, rather than acting like a funnel.

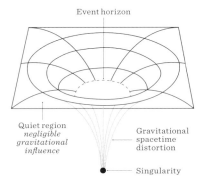

Event horizon

Quiet region
*negligible
gravitational
influence*

Gravitational
spacetime
distortion

Singularity

at 35,000 feet – a hypothesis that has been proven using the most accurate atomic clocks on Earth. The difference is measured in minute fractions of a millisecond, but the difference is measurable nonetheless.

Approach something as massive as a black hole like Sagittarius A*, though, and the effect on the passage of time is far greater. At its event horizon, time is so distorted, it grinds to a halt. Watch an object fall in from outside and you would never see it drop past the event horizon; it would remain frozen in time and only very gradually fade from view. But if it was you falling into Sagittarius A*, your experience of time would be normal, you would see the Universe behind you, in fast forward, before you pass the event horizon and head into the dark abyss. These are two equally real but very different versions of the same event, so mind-bending that for many theoretical physicists it was difficult to believe such a phenomenon could actually exist – theoretically possible, but unimaginable in the real Universe.

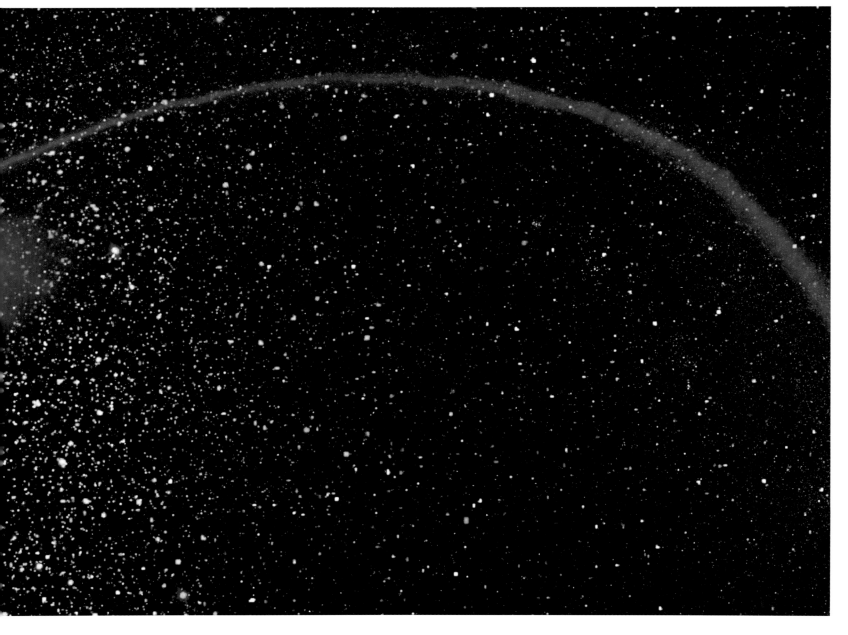

From the unimaginable violence of the sculpting of the galaxy, to the ageing giant that we see today, this is the story of Sagittarius A*, a story that has come together over the last few decades with telescopes like Chandra and Fermi filling in the last few pieces. But we are far from unlocking all the secrets of our galaxy, of supermassive black holes.

Einstein reviewed the fabric of the Universe not as something that is static, but instead as something fluid, something that bends and warps around objects with mass – we call this fluid spacetime, a combination of space and time. Einstein's insight was to realise that these two things are intimately connected to each other. That an object with mass bends space, but also affects the passage of time. It slows time down, and if you are at the event horizon of a massive black hole, time would be so distorted that it would effectively grind to a halt. If you were to

watch a star fall into Sagittarius A*, you wouldn't actually see it cross that event horizon. Instead, you would see it turn redder, freeze and then eventually fade out of sight.

It's not the amount of mass that creates this deep well in spacetime. It's actually how dense this material is. In fact, if you could crush something to high-enough density, anything could become a black hole. If the Sun were compressed into something four miles across, it would be a black hole. If you took the entire Earth and crushed it into the size of a snowball, it would be a black hole out there in the Universe. Most black holes have been formed by the death of collapsing massive stars, but there could never have been a star massive enough to directly form a supermassive black hole. The object that became the young Sagittarius A*, a star, was at least 100,000 times less massive than it is today. It had to grow. Fortunately, food was plentiful.

Relativistic jet
When stars are absorbed by black holes, jets of particles and radiation are blasted out at near light speed. They can extend for thousands of light years into space.

Event horizon
The radius around a singularity where matter and energy cannot escape the black hole's gravity. The point of no return.

Photon sphere
Photons are emitted from hot plasma near the black hole, which bends their trajectory producing a bright ring.

Singularity
The very centre of a black hole where matter has collapsed in a region of infinite density.

Accretion disc
A disc of superheated gas and dust whirls around the black hole at immense speeds, producing electromagnetic radiation (X-rays, optical, infrared and radiowaves) that reveal the black hole's location. Some of this material is doomed to cross the event horizon, but most will be forced out to create jets.

Innermost stable orbit
The inner edge of an accretion disc is the last place that material can orbit safely without the risk of falling past the point of no return.

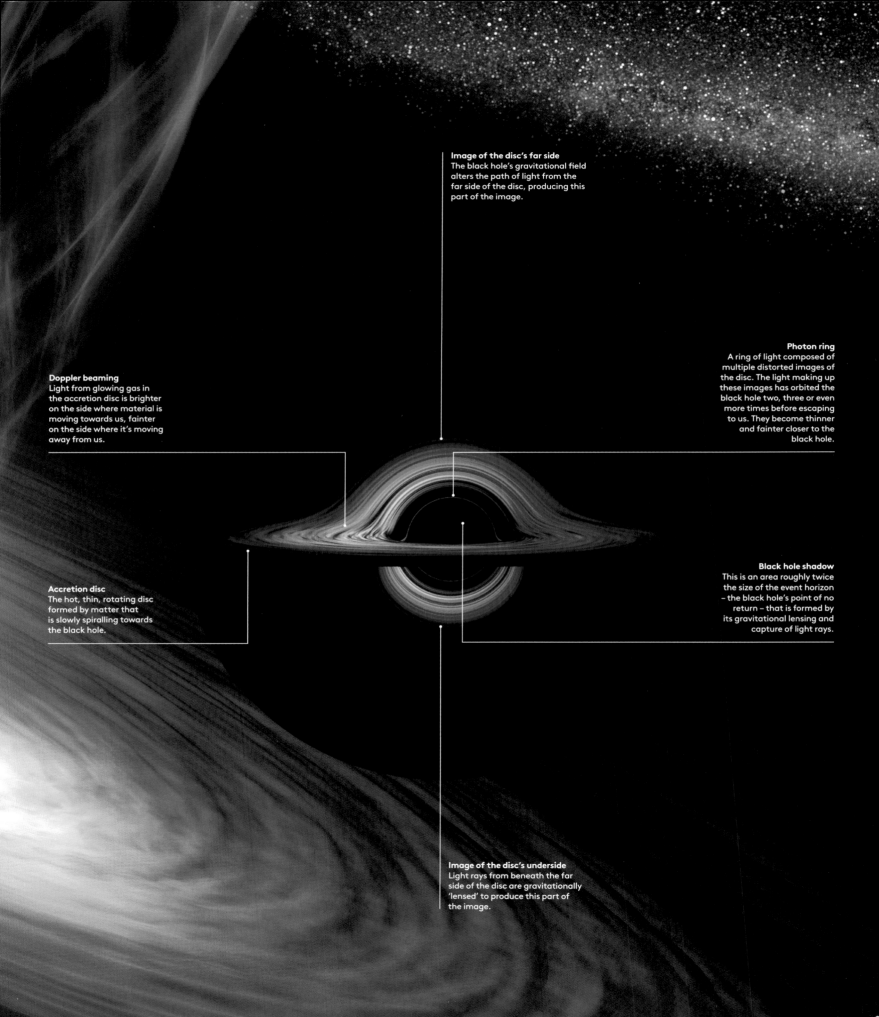

Image of the disc's far side
The black hole's gravitational field alters the path of light from the far side of the disc, producing this part of the image.

Photon ring
A ring of light composed of multiple distorted images of the disc. The light making up these images has orbited the black hole two, three or even more times before escaping to us. They become thinner and fainter closer to the black hole.

Doppler beaming
Light from glowing gas in the accretion disc is brighter on the side where material is moving towards us, fainter on the side where it's moving away from us.

Black hole shadow
This is an area roughly twice the size of the event horizon – the black hole's point of no return – that is formed by its gravitational lensing and capture of light rays.

Accretion disc
The hot, thin, rotating disc formed by matter that is slowly spiralling towards the black hole.

Image of the disc's underside
Light rays from beneath the far side of the disc are gravitationally 'lensed' to produce this part of the image.

SEEING IN THE DARK

Hunting for the darkest objects in the Universe was never going to be easy. Black holes swallow everything that goes near them, including light, making it impossible for astronomers to observe them directly, but that hasn't stopped us. While theoretical physicists painted pictures with their equations to reveal the character of a black hole, astronomers began searching for the evidence of their existence in the clues that were emerging around the edges of these monsters.

The first physical evidence of the potential existence of a black hole came in 1964 with the discovery of a massive celestial X-ray source in the constellation of Cygnus. It was detected using Geiger counters that were carried above the atmosphere by exploratory suborbital rockets launched from the White Sands Missile Range in New Mexico. X-rays of celestial origin are not detectable here on Earth because they are blocked by the atmosphere, but aboard the Aerobee rockets an X-ray map of the sky was produced for the first time.

Cygnus X-1, as it came to be known, is one of the strongest X-ray sources ever observed and its discovery suggested the existence of a celestial phenomenon that was superheating gas to temperatures in the millions of degrees. The exact source of the X-ray emission was not immediately clear and in 1974 Stephen Hawking famously bet against it being a black hole in a wager with his friend and colleague Kip Thorne. It would not be until 1990 that Hawking acknowledged that the evidence (even though it was indirect) had become overwhelmingly in favour of Cygnus X-1 being a black hole and conceded the bet. We now believe Cygnus X-1 to be a black hole 21 times the mass of the Sun, with a huge blue star called HDE 226868 circling around its centre every 5.6 days. Originally, this star system would have been made up of two such stars orbiting each other in a gravitational frenzy, but one of the stars burned brighter and quicker, and speeding through its life cycle it began to expand, shedding its mass as the core burnt through to the end of its fuel.

At this point, rather than ending in a massive supernova explosion before collapsing, we now think that the core of the star collapsed directly, forming what is known as a stellar mass black hole. Pulling in gas from its neighbouring star, Cygnus X-1 feeds on its massive neighbour, superheating the gas to millions of degrees and in doing so creating the X-ray bursts that have allowed us to peer across the Milky Way and study this dark giant.

Cygnus X-1 is perhaps the most studied X-ray source ever observed in the night sky, but the type of black hole, a stellar mass black hole, that it is believed to be coming from is far from a rarity. With a mass 10 to 24 times that of our sun, this type of black hole is thought to occur with surprising regularity. Difficult to detect directly, we can estimate their frequency from the number of stars that are large enough to produce them, and that quick and dirty calculation suggests there may be as many as a billion such black holes in the Milky Way galaxy alone. The galaxy may also be populated by two other varieties of black hole – miniature black holes and intermediate mass black holes. The evidence for the former is entirely theoretical and the existence of the latter is still hotly debated, with many researchers convinced that a class of these intermediate black holes must exist with masses up to 100,000 times that of our sun.

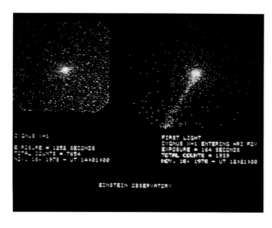

CYGNUS

Left: Cygnus X-1 was the first object seen by the High Energy Astronomy Observatory (HEAO)-2/ Einstein Observatory in 1980.

Opposite: NASA launched a number of key missiles from White Sands Missile Range in New Mexico. Photographed here in 1963.

Right: Two black holes orbiting each other in a combined accretion disc. Eventually they will merge, causing gravitational waves.

X-RAY BINARY STAR SYSTEMS
Material transferred from a star to an orbiting collapsed stellar
core, such as a black hole, forms an accretion disc. The collapsed
companion stars are so dense that the high rotation speeds in
the discs produce X-rays. The gravitational mid-point is marked
by the figure-of-eight loops, the Roche lobes.

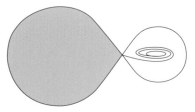

Cygnus X-1

When it comes to the largest of black holes, however, the search is far more limited in number. Supermassive black holes are much rarer beasts that we believe lie at the centre of almost every sizeable galaxy, including our own. But something with the mass of a supermassive black hole, often millions or even billions of times bigger than our sun, doesn't just come into existence with the collapse of a single star; there's never been a star massive enough to create such a monster. To become that big requires a life story that takes a very different course.

While today Sagittarius A* is supermassive, 13 billion years ago, not long after its birth, it appears it was a far smaller beast. Born into a densely packed neighbourhood of youthful stars, Sagittarius A* was a fraction of the diameter of the stars around it – but looks can be deceiving. Although small in scale, its enormous mass meant that this newly born black hole's intense gravitational pull could start preying on

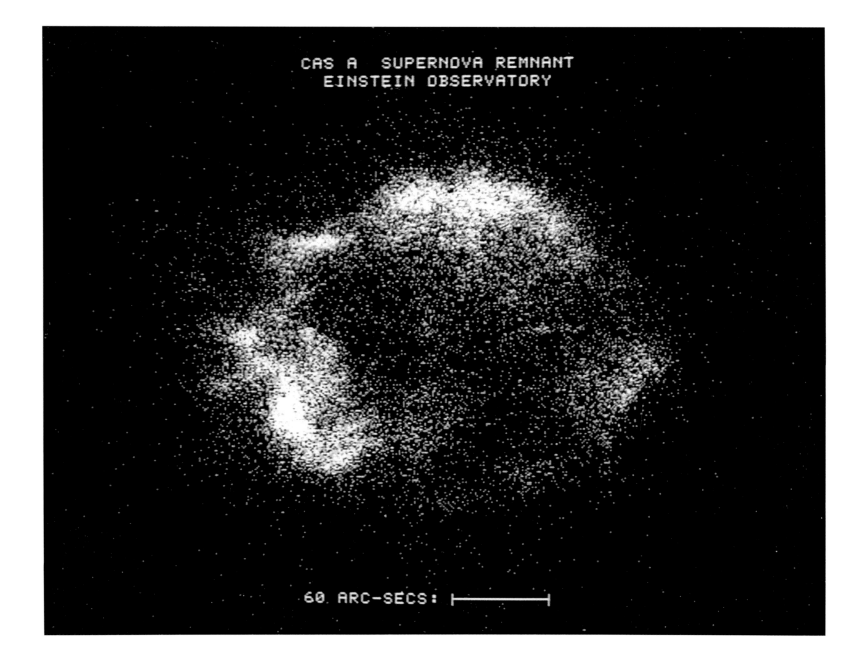

Below: In 2015, gravitational waves caused by colliding black holes were detected by LIGO, 100 years after Einstein's prediction.

Bottom: A technician inspects one of LIGO's core optics (mirrors) by illuminating its surface with light at glancing angles.

Opposite: An X-ray image of the supernova remnant Cassiopeia A shows the expanding shell and a shockwave beyond that.

anything in its immediate vicinity. To begin with, that would have been just the gas and dust that is drawn towards a black hole, slowly feeding the monster with any scraps that drift too close until they have no chance of escape. But that would have been just the beginning. For this black hole's appetite would have rapidly escalated to start feasting not just on dust and gas but on any stars that strayed too close, ripping them apart into a stream of matter that would feed the monster within. This is how we think black holes grow, by consuming the stuff around them, grabbing whatever is within their gravitational reach and pulling it inwards into the darkness.

But even gorging on stars and planets with a few asteroids thrown in for good measure, we still don't think it's possible for a small black hole to become supermassive – like Sagittarius A*. There isn't enough matter to allow a black hole to grow big enough and quickly enough on a diet of stars and stardust alone. Understanding how a relatively small black hole becomes supermassive has long been a mystery, but by searching for a strange phenomenon that was first predicted by Einstein back in 1916, we are finally beginning to find an answer.

The potential existence of gravitational waves emerged as a strange prediction from Einstein's Theory of General Relativity. These ripples in spacetime would in theory be created by the most violent and powerful events in the Universe, sending waves of undulating spacetime travelling out at the speed of light in all directions away from the event. Find one of these gravity waves and within it would be clues to its origin, information that would allow us to delve into and understand the rarest of cosmological events. Over the next 100 years, astronomers searched for evidence of gravitational waves, but detecting an invisible wave was never going to be easy. By the time one of these ripples reaches Earth, they are minuscule disruptions to the fabric of the Universe, wobbling spacetime with a wave hundreds of times smaller than the size of an atomic nucleus. Indirect evidence of Einstein's gravity waves slowly emerged from the study of rare star systems where twin pulsars orbited each other, but their direct detection remained beyond us until well into the twenty-first century.

It would take the building of one of the most ambitious observatories in the history of science to finally make a breakthrough. The Laser Interferometer Gravitational Wave Observatory (LIGO) consists of two observatories built 3,000 kilometres apart in Washington State and Louisiana. Each LIGO detector consists of two arms, each 4 kilometres long, containing lasers that are so precisely calibrated they are designed to pick up the minutest ripple in the fabric of the Universe. After years of refinement and what at times looked like failure, LIGO detected its first gravity wave in September 2015 in an event that allowed us to peer 1.3 billion light years away and witness the collision of two black holes for the very first time. Since then, LIGO and its European counterpart, Virgo, have been able to make multiple observations of the events that create gravitational waves, and many of these involve black holes colliding.

Black holes, it seems, are far from lonesome monsters. LIGO had provided compelling evidence to suggest that black holes grow not just by consuming stardust but by consuming other black holes, and this has led us to be able to paint a new picture of how Sagittarius A* may have evolved from newly born to supermassive black hole.

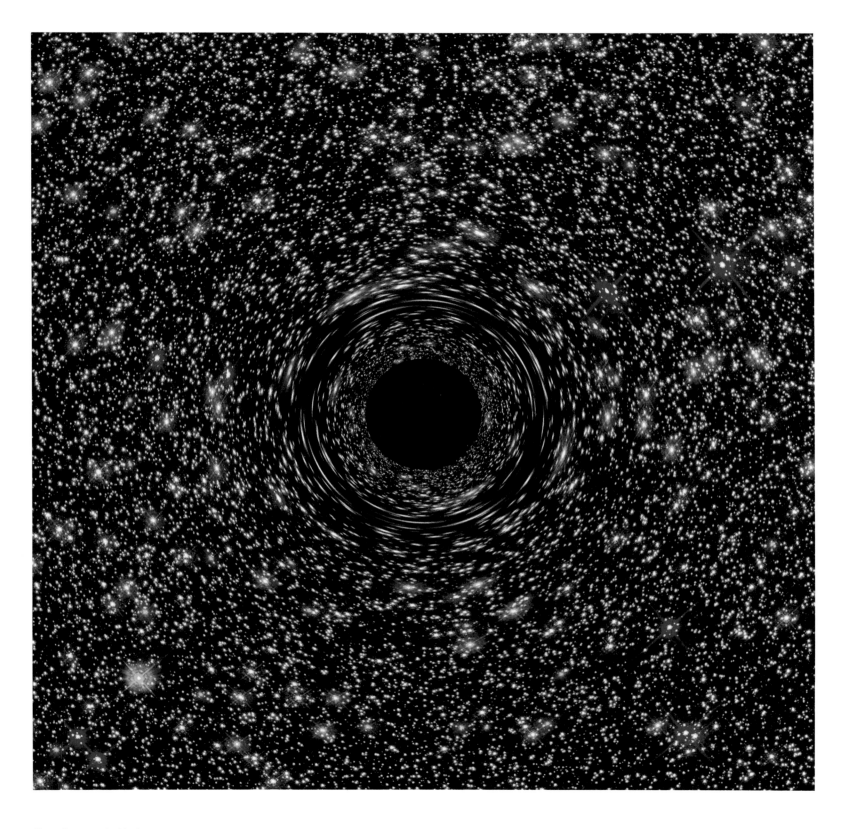

Above: Supermassive black hole at the core of a galaxy, where its powerful gravity distorts space around it like a funhouse mirror.

Creators of our galaxy

I work on supermassive black holes – black holes whose masses are about a billion times greater than our sun. That sounds like a lot, but a supermassive black hole is actually physically a relatively small object compared to galaxies. The event horizon of a billion solar mass black hole would be about the size of our solar system. And if you compare that to the massive galaxy that black hole is sitting in, it's like comparing the size of a grape to the size of the Earth. We're talking a billion-fold contrast in scale. And yet that tiny little grape at the heart of that Earth-sized galaxy can absolutely sculpt the evolution of the galaxy that it's sitting in. So over thousands and thousands and thousands of light years, this solar-system-sized object is sculpting the evolution of gas.

The gravitational potential energy that you must liberate in order to grow a billion solar mass black hole exceeds the binding energy of entire galaxies by about a factor of one hundred. So, if even a fraction of that energy that must be liberated and growing that black hole couples to the ambient environment of the galaxy, then the black hole itself can have a really profound, even dominant impact on the evolution of the entire galaxy. We think black holes are the orchestrators of the cosmic symphony in many, many ways, and they do so through a process called black hole feedback. They liberate so much energy, and that energy has got to go somewhere. It turns out that energy is the ultimate driver of many of the major aspects of galaxy evolution.
Grant Tremblay, Astrophysicist, Harvard Smithsonian Center

When it was born, Sagittarius A* was not just surrounded by stars, there were kindred spirits, too – other black holes circling in its vicinity. At some point, it is believed that another black hole came close enough to the young Sagittarius A* to become gravitationally entwined, pulling on one another in a courtship display of gravitational might that locked them into an inescapable spin. A dance that would only end one way, with these two dark stars hurtling towards each other at half the speed of light.

Spiralling into each other in a collision of unimaginable force, the resulting merger would have left Sagittarius A* bigger and more fearsome than ever, with perhaps double the mass. The newly emboldened black hole's gravitational pull would have been even stronger, allowing it to pull in ever more mass from the surrounding stars and dust clouds and perhaps undergoing multiple collisions with other black holes, as it became set on its path to becoming the dominant force at the centre of our galaxy.

The merger that set Sagittarius A* on its path to becoming the behemoth of today happened many billions of years ago, with all traces of the ripples created by the black hole mergers long gone. But since the first breakthrough by LIGO in 2015, we have been detecting the ripples of spacetime created by other merging black holes across the Universe, none larger than the signal detected in May 2019 that we have named GW190521. In an event that astronomers have described as 'probably the largest explosion we've ever known in the Universe', two goliaths smashed into each other – one 85 times the mass of the Sun, the other 66 times bigger, with the resultant black hole 142 times as massive as the Sun. A quick bit of mental arithmetic will tell you that leaves nine solar masses unaccounted for, but this mass didn't just disappear, it resonated out through the Universe in the form of energy, gravitational waves that carried the echo of this event for over 7 billion years, before reaching our planet, and the technology we had built that was waiting to listen to such an event, on the morning of 21 May 2019.

We are still grappling with all the consequences of this extraordinary observation, but this and other discoveries like it are starting to help fill in the gaps in our understanding of how black holes evolve. From giving us the first indication of the existence of intermediate-sized black holes, to the realisation that the evolution of black holes may involve multiple mergers across billions of years as some like Sagittarius A* grow from stellar mass black holes, through a number of intermediate stages on the way to supermassive status.

By the time Sagittarius A* had fed its way to adulthood, it had not only transformed itself but also the galaxy around it. With its mass and power growing, it was able to attract ever more stars and gas into its gravitational hold, not close enough to be consumed but instead held in a slow, spinning orbit. That would transform an unshaped collection of dwarf galaxies and star clusters into a spiral galaxy, the majestic Milky Way with Sagittarius A* at its core, many tens or even hundreds of thousands of times more massive than any star in the Universe. And Sagittarius A* is not unique; we now think that virtually every large galaxy has a supermassive black hole at its heart, and in the last couple of years we have finally managed to glimpse what one of these beasts actually looks like.

PEERING INTO THE ABYSS

This remarkable image is our first of a black hole. A photograph that until its release in April 2019 was thought to be impossible to take. It took the work of a team of international scientists over a decade to capture the silhouette of this supermassive black hole, 53 million light years from Earth, 6.5 billion times the mass of the Sun and sitting at the centre of a galaxy called M87. To take it required the creation of a telescope almost as big as the planet itself. Known as the Event Horizon Telescope (or EHT), this global network of multiple ground-based radio telescopes work together to create one vast virtual telescope with the power to peer into the furthest reaches of the Universe.

The EHT uses a technique called very-long-baseline interferometry, which synchronises an array of eight smaller telescopes to focus simultaneously on the same object at the same time. The effective aperture of this virtual telescope is equivalent to the distance between the two farthest contributing bases, and so in the case of the EHT that is the distance between the South Pole Telescope at the Amundsen-Scott research station in Antarctica and the IRAM Telescope in the Sierra Nevada, in Spain. With an aperture approaching the diameter of the Earth, the EHT is able to peer further into the darkness, capturing more light at a higher resolution than any single telescope on our planet. Launched back in 2009, this massive international collaboration began with two primary targets for the telescope. The first was Sagittarius A*; at just 26,000 light years from Earth this would appear to be the obvious first choice, but despite its size and proximity our view of it is blocked by the gas and dust that orbits the galactic centre, making any image obscured by this pollution, even with the use of radio astronomy.

THE EVENT HORIZON TELESCOPE (EHT)
The EHT is a large telescope array consisting of a global network of radio telescopes spanning the globe, from Hawaii to North America, South America, Europe and Antarctica.

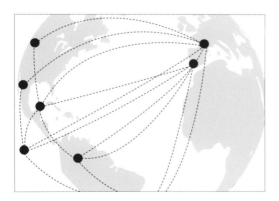

Right: The first image ever taken of a black hole. This is an EHT image of the supermassive black hole in the centre of the galaxy M87.

Opposite: The Atacama Large Millimetre/submillimetre Array (ALMA) in the Atacama Desert is a key part of the EHT Telescope.

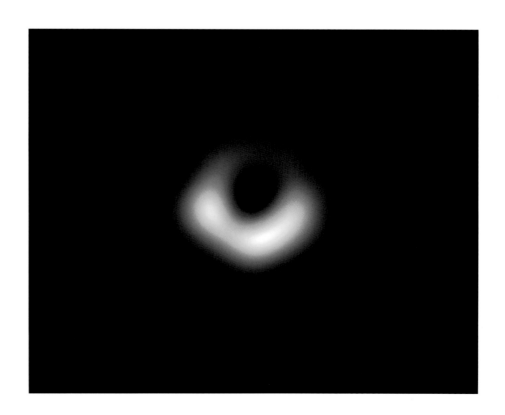

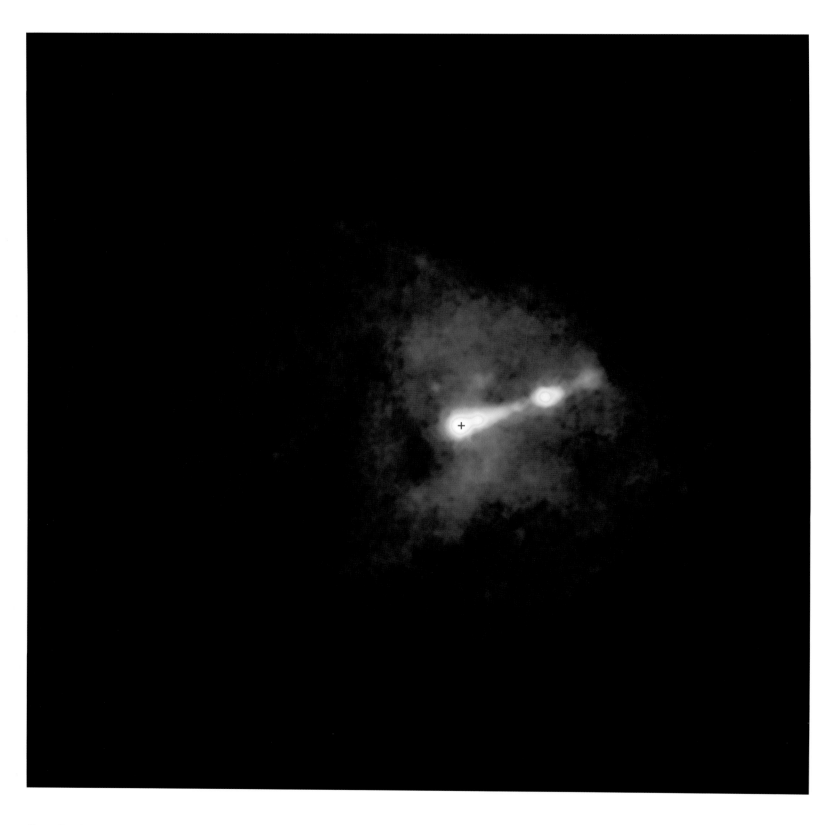

Above: Chandra studies the M87 black hole in combination with the EHT radio image to learn more about its event horizon.

Peering into darkness

Even with the Event Horizon Telescope, we cannot directly observe a black hole because it is literally shielded from us. A black hole is shrouded, by definition, by an event horizon, a causal discontinuity in spacetime. It is really, truly black. But what we can observe is the profound effect of matter around black holes. One of the ways in which we do that is with the Event Horizon Telescope itself, this photon ring of light, the sort of last orbit of doomed gas spiralling around the event horizon.

The thing that still really blows my mind about black holes is that they warp not only space, but time itself. That turns out to be true of all matter. I'm warping space and time around me right now. You're doing the same thing. All mass does that. But black holes do it in a really extreme way, such that if you were to be able to be in an impossible spaceship and shielded from radiation and surviving around the orbit of a black hole, and if you were to take your pocket watch with its minute hand going around the clock and drop it towards the black hole as it sunk away towards the event horizon, you'd see the second hand of that clock ticks slower and slower and slower. And the moment that it reached the event horizon, you would see the second hand on that clock stop. Time literally stops at the surface of the event horizon. Things are frozen there for all of eternity.
Grant Tremblay, Astrophysicist, Harvard Smithsonian Center

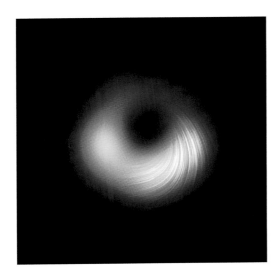

Above: The EHT's M87 image shown in polarised light. The lines on the image mark the orientation of polarisation of magnetic fields.

Second on the list was the vast elliptical galaxy Messier 87 – or M87 for short. This galaxy was thought to contain one of the largest supermassive black holes ever discovered, substantially larger than Sagittarius A*. M87 was also a prime target due to the fact that it was also thought to be highly active, drawing in matter that would in theory create an even brighter halo for the EHT to image.

With each of the telescopes synchronised together with atomic precision, the EHT team chose a ten-day window in April 2017 in which it would point towards the heart of M87 with a telescope that had an effective resolution 4,000 times greater than the Hubble Telescope. With a large dash of luck, the chosen time period coincided with four clear days across all eight of the observation sites, and so the EHT was able to capture an unparalleled amount of data, 5 petabytes of information that could then be crunched by two independent teams in the USA and Germany.

The result is one of the most historic cosmological images we have ever taken. An image that reveals more in what we cannot see than in what we can. A silhouette of a supermassive black hole, its form revealed by the presence of a lopsided ring of light, the superheated glow of matter falling into the abyss.

We can't directly see the black hole, of course, as the event horizon is lost in the impenetrable darkness, somewhere towards the centre of the image. But what we can see is the bright disc around the black hole, known as the accretion disc, the outlying ring of glowing gas that surrounds the event horizon. Look at that image and you are looking at a black hole 6.5 million times the mass of the Sun, with huge swathes of gas and dust spiralling into it, heating up and emitting large amounts of radiation that has sped across the Universe, enabling us to photograph it from a distance of 318 quintillion miles.

Since the start of writing this book, the historic first image of M87 has already been updated. In March 2021, the team behind the Event Horizon Telescope released a new image that added polarised light to the picture. As well as adding a subtle beauty to the image, the polarised light also gives us the first indication of how magnetic fields behave around a supermassive black hole. Fifty times the strength of the Earth's magnetic field, it's clear from the swirling spiral pattern of the light that the magnetic field is ordered. This is important because it begins to help us explain how supermassive black holes like M87 can release powerful jets of matter out into the surrounding galaxy, jets that are launched and guided by magnetic fields.

In front of our very eyes, we are seeing not just the might of these extraordinary monsters but also the guiding hand that every supermassive black hole gives to its home galaxy. Sending out vast jets of matter, they are shaping it, nurturing it and ultimately driving the evolution of not just the galaxy but every star and planet within it. By interacting with their environment, and with the extreme energies involved, supermassive black holes can influence the evolution and fate of entire galaxies, including our own.

CLOSER TO HOME

The story of Sagittarius A* may stretch back 13 billion years or so, but the story of our understanding of the black hole at the centre of our galaxy begins much more recently than that. It began with the work of Karl Jansky, an American engineer and pioneer of radio astronomy who was working at the Bell Telephone Laboratories in the late 1920s when he made an unexpected discovery. Using a massive 30-metre-wide antenna that he'd built to detect radiowaves at a frequency of 20.5 Hz, Jansky began sweeping the sky for signals, turning the antenna on a set of four Ford Model-T tyres so that he could pinpoint the direction of any signal. After months of observation, Jansky had become intrigued by a mysterious signal he could not explain, 'a steady hiss-type static of unknown origin'. Almost all of the signals that the antenna picked up could be identified as static coming from thunderstorms, both near and far, but the faint steady hiss that peaked once a day could not be explained so simply.

Jansky spent a good part of the next year exploring this mysterious signal, watching as its peak position drifted across the sky in a cycle that repeated in the familiar pattern of 23 hours and 56 minutes. This period of time is known as the sidereal day – an astronomical timescale that relates to the Earth's rotation relative to the stars rather than the Sun. By cross-referencing his findings with astronomical charts, Jansky was able to conclude that the signal seemed to be emanating from deep within the centre of the Milky Way, towards the constellation of Sagittarius. Publishing his findings in 1933 in a paper entitled 'Electrical Disturbances

Left: Karl Jansky with his antenna, collecting data that would herald the birth of radio astronomy.

Below: Astronomical clock showing the right ascension of the Sun and Moon. The two lower dials show mean solar and mean sidereal time.

Bottom: Thunderstorms emit static electricity, which registers on radio telescopes.

Apparently of Extraterrestrial Origin', the discovery was widely publicised in the popular press as a possible signal from an extra-terrestrial intelligence. Jansky dismissed such speculation and wanted to explore the mysterious 'star noise' in more detail, but caught between a dismissive astronomical community (Jansky was an engineer, not a 'scientist') and an employer who was more interested in trans-Atlantic communications than unexplained astronomical signals, Jansky received little support to move his discovery forwards. Instead, having listened in on the dark secret at the centre of our galaxy, Jansky's discovery would not be fully built on for another 40 years.

It wasn't until 1974 that the 'steady hiss of unknown origin' discovered by Jansky began to reveal more of its secrets. Working for the National Radio Astronomy Observatory, astronomers Bruce Balick and Robert Brown began investigating the mysterious radio source using the Green Bank 35-kilometre radio link interferometer based in West Virginia. With the advancements in radio astronomy, Balick and Brown were able to observe that the signal emanating from Sagittarius A* consisted of a number of different overlapping components, but one in particular stood out. On 13 and 15 February 1974, the two astronomers discovered a small core in the galactic centre that was emitting radiowaves far more brightly than anything else. They went on to name this high-intensity radio source Sagittarius A*, using the asterisk to reflect the 'excitement' of the discovery, a symbol that is normally used by physicists to signify an electronically 'excited state' when describing an atom.

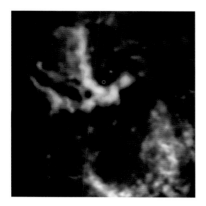

We were finally beginning to home in on the identity of the mysterious object at the centre of our galaxy. With a more clearly defined target in their sights, astronomers were now able to begin exploring the source of Sagittarius A* and start attempting to understand its characteristics. In the late 1970s, a team from the University of California, Berkeley, began using infrared telescopes to peer through the dust that was obscuring the galactic centre. Measuring the velocity of swirling clouds of ionised neon gas, they were able to estimate the mass around which this gas (and a large cluster of stars) seemed to be orbiting. The extraordinary result of their calculations pointed in one direction only – contained within a tiny space in the centre of our galaxy was a mass equal to approximately 3 million suns (now known to be 4.3 million solar masses). Such a large mass contained within such a small space strongly suggested that the radio source Sagittarius A* was being generated by a supermassive black hole. But despite the growing weight of evidence, it would take a Nobel Prize-winning piece of research to finally cross the evidence threshold and lay bare the real identity of Sagittarius A* once and for all.

'Black holes collect problems faster
than they collect matter.'
Carl Sagan

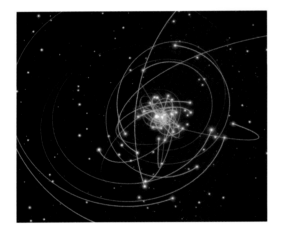

Over a ten-year period starting in the early 1990s, an international team of
astronomers, led by Reinhard Genzel at the Max Planck Institute for Extraterrestrial
Physics in Germany, and Andrea Ghez of the University of California in Los Angeles,
began studying the motion of a collection of stars that orbited around the centre
of our galaxy, just a few light years away from its massive dark heart. By studying
the orbits of these stars, they hoped to be able to tell if there was a single object,
a supermassive black hole that these stars were tightly orbiting, or a collection of
objects – stellar remnants, neutron stars or even small black holes making up the
vast mass of Sagittarius A*. In principle it sounds simple, but in practice tracking
these stars deep in the heart of the Milky Way is not easy. Ground-based telescopes
had to overcome the blurring effect of the Earth's atmosphere, making accurate
tracking difficult, and space telescopes were not suited to the long observation
times required to study the orbits of these stars.

Using the European Southern Observatory's infrared telescope in Chile and
the Keck Telescope in Hawaii, the team spent years developing techniques to try
to overcome the technical challenges of accurately tracking these stars, but it
wasn't until the introduction of a system called 'Laser Guide Star Adaptive Optics'
that the real breakthrough came. By firing lasers into the night sky this technique
allows the creation of reference points, 'artificial stars' generated by the powerful
lasers that can be used to correct the blurring effects of the Earth's atmosphere.
Introduced in 2004, this new technology allowed the team to observe dozens of
stars within just 0.1 light years of Sagittarius A*, known as the S stars, and one of
them – S2 – became of particular interest. S2 orbits Sagittarius A* every 16 years
within just 120 astronomical units (17,951,744,000 kilometres) of the vast mass
that we now know is a supermassive black hole. Reaching speeds of up to 5,000
kilometres per second or one-sixtieth of the speed of light, it is the fastest-orbiting
object we have observed in the Universe.

By 2008, the team had been able to study a complete orbit of S2 and 28 other S
stars by measuring their distances, speeds and orbits with such precision that they
were able to calculate that Sagittarius A* had a mass of 4.31 (+/-0.38) million suns
within a space no bigger than the size of our solar system. Announcing the results
that would go on to win Genzel and Ghez the Nobel Prize in Physics in 2020, they
published the empirical evidence that finally proved that supermassive black holes
really do exist. Commenting on the publication in the *Astrophysical Journal* in
2009, Genzel concluded that 'The stellar orbits in the Galactic Center show that
the central mass concentration of four million solar masses must be a black hole,
beyond any reasonable doubt.' Stranger than fiction, black holes have gone from
theoretical anomalies at the beginning of the twentieth century to observable facts
by the beginning of the twenty-first century. With this discovery, the age of black
hole exploration had truly begun.

CREATION STORY

'Ignorance is never
better than knowledge.'
Enrico Fermi

Ten billion years ago, as our primitive galaxy continued to develop around the dark force at its centre, a new age for our black hole was on the verge of beginning. This was a time of abundance in the Milky Way's history, a galaxy brimming with gas and dust, providing an endless offering of sustenance for Sagittarius A* to feast on and grow. But for all of its gluttony, Sagittarius A* was not just an agent of destruction.

Creation and destruction often go hand in hand in the Universe, and black holes are no exception. As the black hole sucked vast amounts of cosmic matter into its gravitational grasp, not all of it disappeared across the event horizon. Much of it stayed on the outside, forming a swirling cloud that orbited the black hole in a pattern that would almost certainly have mirrored the images we have seen of the black hole at the centre of M87. And just like M87, the halo around Sagittarius A* would have been far from peaceful. During intense periods of activity, the halo would have been violent and volatile and the powerful magnetic fields that swirled around it would have been able to throw out great swathes of this material. Ejecting it out along the magnetic poles of the black hole, creating vast jets that would have swooped through the galaxy, today this halo lies quiet and Sagittarius A* lies dormant, but we know from our observations of other galaxies that supermassive black holes rarely stay quiet for long. Even in its recent past, Sagittarius A* would have gone through periods of far more intense activity, ingesting the large amounts of matter that we know can power the high-energy jets we have witnessed emanating from other supermassive black holes. We even think we might have now seen the echo of these powerful jets in an extraordinary structure within our galaxy that we have only recently discovered.

Launched on 11 June 2008, the Fermi Gamma-ray Space Telescope entered into low Earth orbit and began its mission to explore some of the most mysterious and powerful phenomena in the Universe. From neutron stars to supermassive black holes, Fermi was built to detect the traces of objects that can generate inconceivable amounts of energy in the form of gamma-ray radiation. These are the most energetic wavelengths of light, billions of times more powerful than the visible light we can see with our eyes. As Fermi set to work, it allowed us to begin exploring through gamma rays parts of the cosmos that normally hide in the shadows.

One of Fermi's primary mission objectives was to explore the sources of gamma rays within the Milky Way. Unseen to the human eye, our galaxy glows bright with gamma-ray radiation. Generated by cosmic ray particles smashing into the interstellar dust and gas, these interactions create three-quarters of the gamma rays in our galaxy. With the launch of Fermi, the hope was that we would be able to explore the structure, composition and characteristics of the interstellar medium that fills our home galaxy and produces this endless glow.

Left: Spectacular jets powered by the gravitational energy of a supermassive black hole in the core of the elliptical galaxy Hercules A.

Right: The Fermi Gamma-ray Space Telescope in the nose cone of the Delta II rocket launch vehicle.

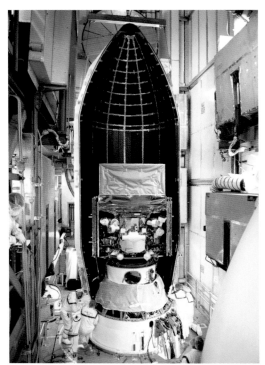

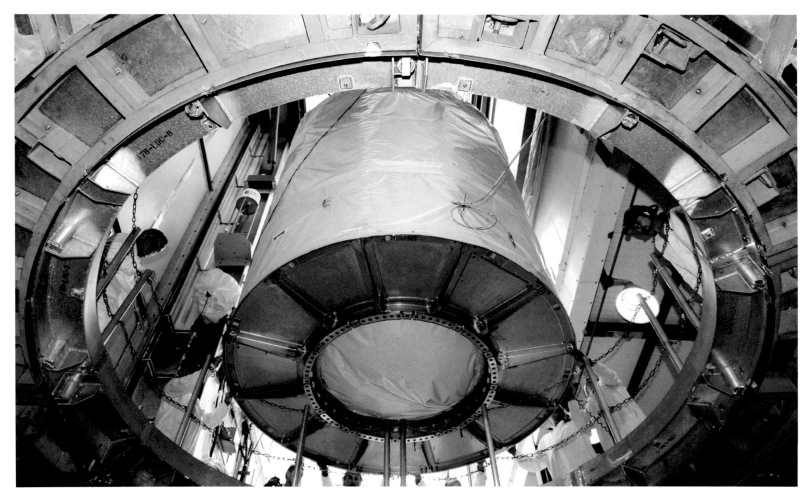

Above: Fermi launch preparations in progress. Here the second stage of the Delta II rocket launch vehicle is put into place.

Top right: Fermi's LAT (silvery box at the top) is shown integrated with the spacecraft systems in December 2006.

Opposite: Within 50,000 light year Fermi bubbles, energetic electrons are interacting with lower-energy light to create gamma rays.

THE FERMI LARGE AREA TELESCOPE (LAT)

Cross-section of the LAT showing how conversion foils transform an incoming gamma ray into a pair of particles – an electron and a positron. The LAT tracks these particles as they travel through the instrument, reconstructing the gamma ray's original direction. The particles deposit their energy in the Calorimeter, which measures the gamma ray's energy.

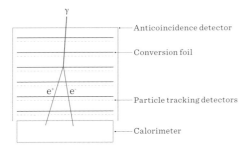

But as Fermi orbited the Earth and began constructing a map of the Milky Way's gamma-ray glow, it saw something that scientists could hardly believe, an astonishing discovery likened to finding a brand-new continent here on Earth. Emerging from the plane of the Milky Way were two enormous structures, bubbles that each stretched out over half the width of the galaxy itself. These bubbles of superheated gas reach temperatures in excess of 10 million degrees as they stretch out 26,000 light years from the plane of the galaxy. They are so large that if our eyes were sensitive to the gamma-ray wavelengths of light emitted by the bubbles, they would fill half the night sky from our vantage point here on Earth.

To make a structure as expansive and hot as this requires enormous amounts of energy; the event that created what have become known as the Fermi bubbles is still in part a mystery, but where they came from is not. The orientation of the bubbles extending symmetrically above and below the galactic core strongly suggests that their origin is tied to events that took place around its own supermassive black hole. And we have seen from that landmark first image that black holes like M87 can not only pull in matter but also accelerate out jets of gas at unimaginable speeds. So, what if Sagittarius A* had done exactly the same in the distant past?

Today it sits relatively quietly, but we know this has not always been the case. Over the last 12 billion years there have been times when our black hole, like every black hole, has got a little too greedy, and instead of consuming all the matter it pulled inwards, it heated it up and flung it back out, and we now think the Fermi bubbles are the remnants of just such an event. Driven by a complex dance of

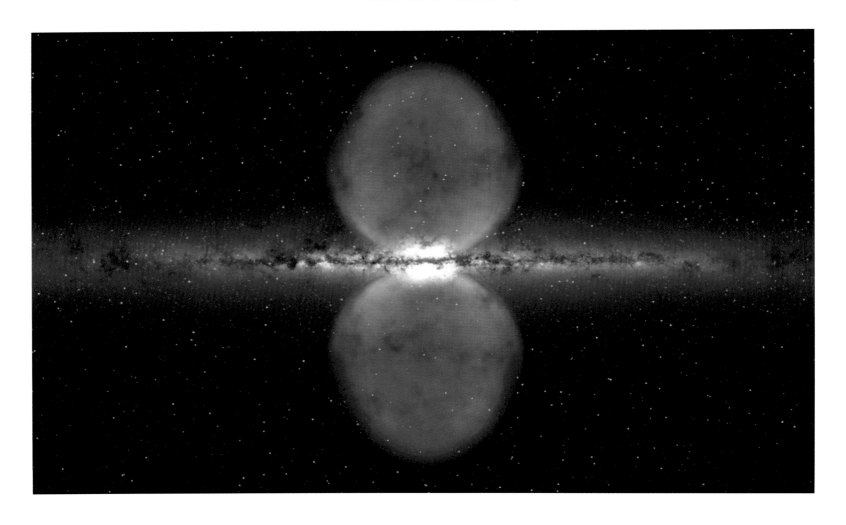

ACTIVE GALACTIC NUCLEUS (AGN)
'Active galaxies' have an AGN that emits a range of non-stellar
radiation, including radio, microwave, infrared, optical, ultraviolet, X-ray and gamma ray radiation. The most powerful AGN
are classified as quasars.

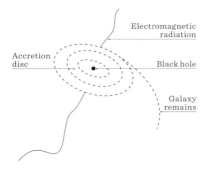

electrical and magnetic forces around the black hole, Sagittarius A* sent vast
amounts of matter outwards in the form of superheated jets of gas travelling near
to the speed of light trillions of miles into space. The remnants of that intensely
energetic event are still visible to this day in the vast bubbles that the Fermi
telescope discovered. In fact, we expect that such an event may not have been a
one-off but has happened multiple times, fundamentally changing the structure
and nature of our galaxy and having a profound effect on its evolution.

These gigantic features offer clues that our black hole isn't solely an agent
of destruction; in fact they suggest the opposite, that it may also have been the
ultimate galactic creator. Driven by bursts of brief but intense activity, lasting for
perhaps just a few tens of millions of years, we think Sagittarius A* has been firing
off like this periodically throughout its life, blasting hot gas once trapped in the
galaxy's centre trillions of miles into space, that then fell like cosmic rain onto the
outer reaches of the Milky Way. Inject such a huge amount of energy and matter
into a system and one thing is certain, it's going to have consequences.

Recently, a remarkable idea has started to emerge that links these enormous
outpourings of energy to the emergence of our solar system and the Earth as a world
upon which life could not only begin but thrive. There are, of course, an enormous
number of interconnecting factors as to why life not only began here on Earth but
was also able to develop in an unbroken line for 4 billion years, allowing it to evolve
into the complex living world that we see today. And at first glance it seems difficult
to believe that one of those factors could be the activity, billions of years ago, of a
supermassive black hole at the centre of our galaxy. But we are now beginning to
suspect that those great outpourings of energy from Sagittarius A* played a crucial
role in making our region of the Milky Way one in which life could flourish.

In the deep and distant past, as the hot gas released by Sagittarius A* rained down
on this corner of the galaxy, it not only delivered an abundance of the heavy elements
that are crucial to build our star and planet but it may also have had a dampening
effect on the rate of star formation in this part of the Milky Way. Now, you might
think that a hot gas cloud would produce more stars, but in fact the opposite is true,
because heat means that everything's moving around very fast and that makes it
more difficult for gravity to grab hold of the gas clouds and collapse it to form stars.

So we think the explosive influence of Sagittarius A* reduced the number of
stars that formed in this region of the galaxy, clearing the way for the most precious
factor in our planet's history to prevail – stability. By rights our corner of the galaxy
should be a violent, inhospitable place filled with giant stars, supernovae and
angry red dwarfs, but Sagittarius A*'s violent outbursts changed all of that, turning
what was potentially a violent region of our galaxy into a peaceful one. So here we
have a stable stellar neighbourhood where 4.5 billion years ago around one small,
inconsequential yellow star something remarkable happened, where, safe from the
violence and radiation of exploding stars, life not only began, but it flourished. A
peaceful corner of the cosmos where life forms with grand ambitions could quietly,
slowly and steadily evolve.

But in breathing life into its realm, Sagittarius A* drew the life out of itself. With
much of the gas, dust and stars that once lay close by cleared away, there was little
left for it to feast on. Our black hole's roar across the Milky Way weakened, turning
it from monster to sleeping giant, a brooding beast hiding in the darkness.

Above: Fermi captures
a blazar firing (centre).
The plane of the Milky Way
appears as a curve below.

Opposite top: The distance
that particles in jets travel
before they 'turn on' and
become bright sources of
light is the acceleration zone.

Opposite bottom:
Supermassive black holes
blast radiation and ultra-fast
winds outward, in a nearly
spherical distribution.

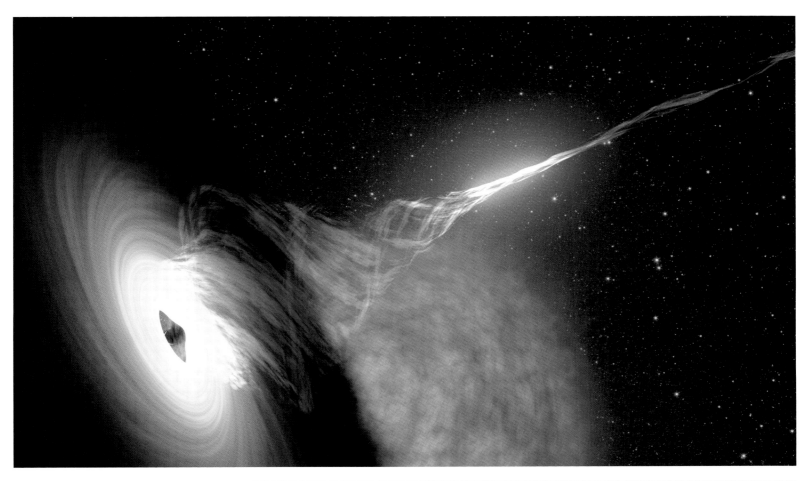

STEP INSIDE

Sagittarius A*'s journey from violent destroyer to sculptor of the galaxy and to the sleeping giant that we see today has been pieced together over the last 20 years by a series of extraordinary advances in the astronomical sciences, led by the observational data from space telescopes such as Chandra and Fermi. It has allowed us to begin understanding in great detail how a supermassive black hole is born, how it grows and how it interacts with its environment. But this is only one side of the story. We have borne witness to a black hole's powerful influence on a galaxy, but the forces it exerts externally are nothing compared to the forces that come into play once you cross over to the other side. Travel beyond the event horizon and into the heart of the darkness and even our best theories are incapable of fully describing the extremes that lie within.

But just as we have illuminated the darkness on the outside, so we are beginning to lift the veil to reveal what is going on inside. And it's by looking to Sagittarius A*'s future that we can begin to comprehend what actually happens when our equations fail us, when infinite forces meet infinite densities, changing the very nature of space and time.

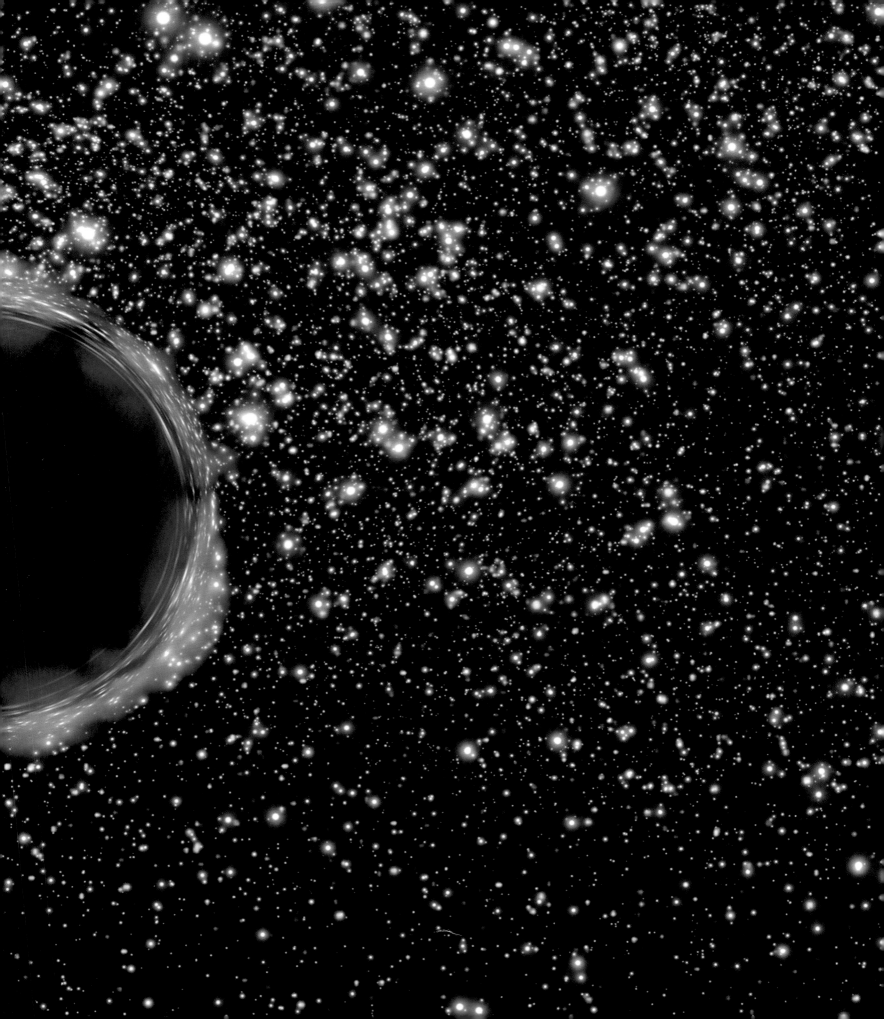

STARFALL

Tucked closer into the black hole at the centre of our galaxy than any other stars that we know of, the S stars are a small cluster of high-velocity stars orbiting Sagittarius A* in a collection of highly eccentric orbits. As we saw earlier in the chapter, the identification and tracking of these stars has been conducted by some of the most sensitive telescopes on Earth, working together to create a planet-sized virtual telescope that has been able to peer into the centre of the galaxy and identify around 20 or so stars in this intense cluster. The orbital behaviour of these stars has been crucial in allowing us to confirm the existence and characteristics of Sagittarius A*, and in particular one of the brightest of these stars, S2, has been observed in great detail. It is this star that we followed on its 16-year orbit of the galactic centre allows us to determine the mass of the object in the galactic centre and therefore to say with confidence that there is a supermassive black hole sitting at the centre of our galaxy.

Located 26,000 light years from Earth, we know relatively little about S2, but current thinking suggests that it is a B-type main sequence star, 10 to 15 times as massive as our sun and relatively young, perhaps just a few hundred million years old. Extremely bright and burning blue, we have little idea where S2 formed or how it came to be so tightly orbiting the black hole. But if it's anything like other stars in our galaxy, then one thing we can surmise is that it almost certainly has planets orbiting it as well, and that opens up some mind-boggling possibilities.

On 19 May 2018, astronomers at the Very Large Telescope in Chile tracked S2 as it made its closest approach to Sagittarius A* on its voyage around the black hole. Coming within touching distance of the vast black hole, at a distance about 120 times greater than the Earth is from the Sun, S2 was subjected to the extremes of

Below: A planet too near to a black hole will be ripped apart because of the much greater gravitational pull on the side closest.

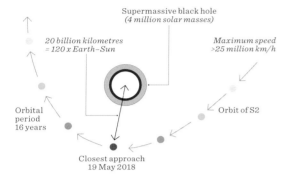

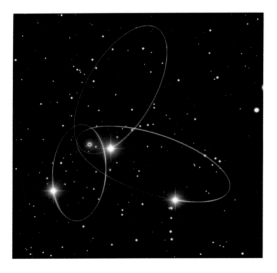

Above: Three stars, including S2,
orbiting very close to Sagittarius A*,
at the centre of the Milky Way.

Sagittarius A*'s magnetic field. Accelerating to speeds of 8,000 kilometres per second, we saw the light of the star redden as predicted by Einstein's Theory of General Relativity, stretched out by the vast gravitational forces. What effect such gravitational forces have on any planets orbiting around S2 is unknown, but what we can imagine is that as S2 is drawn into a closer orbit there will come a time when there will be no chance of escape.

At some point in the future, S2 will drift just a little too close to Sagittarius A*. Pulled into the grasp of the black hole, there will be no escape for this star or any of the planets that are orbiting it. Instead, these worlds would be on a one-way journey down the throat of the beast, a journey that we can now unpick, guided by our theoretical understanding and brought to life in our imaginations.

Let's imagine a rocky planet around S2 is the first to be captured by the black hole's gravity. In front of it is only darkness, but if we could stand on this planet as it descends beyond the point of no return and look back we would see something wondrous. The fierce gravitational forces would not only grip space but time. For us on this planet, time would appear to tick normally, but look back and the Universe would be playing out at astonishing speeds. Millions of years condensed into every second, as if the whole history of the Universe is playing out before our eyes. The life story of a galaxy, its stars, its planets, its life and civilisations would be nothing more than the merest flicker of time. And yet from the outside looking in, we would appear to be slowing in time. Due to an effect known as gravitational time dilation, a prediction from Einstein's Theory of General Relativity, it would appear to observers following the fate of S2 and its planet's descent that time was coming to a standstill on this world, and by taking an infinite amount of time to reach the event horizon but never reaching it, our world would simply fade from view.

But for us standing on the doomed planet, the journey would continue onwards. Crossing the event horizon, we'd feel nothing special; despite appearances, there is no moment of plunging into the darkness, the black hole sucks in light from all directions, allowing us to see out but knowing we can never be seen again. How long this moment of isolation would last we do not know, but we do know that eventually the gravitational forces would become too much. Putting aside the nuances of our own mortal destruction, the ground beneath our feet would slowly be ripped apart, boulders becoming rocks, rocks becoming sand as we plummet towards the infinite forces of the singularity where every journey into a black hole will end. Speeding inwards, the remains of the planet closest to the singularity accelerate to far greater speeds than the trailing remnants, stretching the disintegrating planet across huge distances while simultaneously driving every atom towards what, according to general relativity, is the complete removal of this matter from existence, the ultimate death of not just a world but of every subatomic particle that it is made of.

We once thought this was the destiny of everything in our galaxy. Over trillions of years, the stars and planets around Sagittarius A* would flicker out of existence as they fell inwards, while our supermassive black hole lived on, locking away the contents of every living thing, every planet, every star inside the black hole. According to general relativity, if nothing can ever escape from a black hole, if Sagittarius A* really is an eternal prison, this is the end of the story of the Universe. A darkness littered with holes in spacetime. But this is not the end of the story, because we believe even black holes have a lifetime – even Sagittarius A* will die.

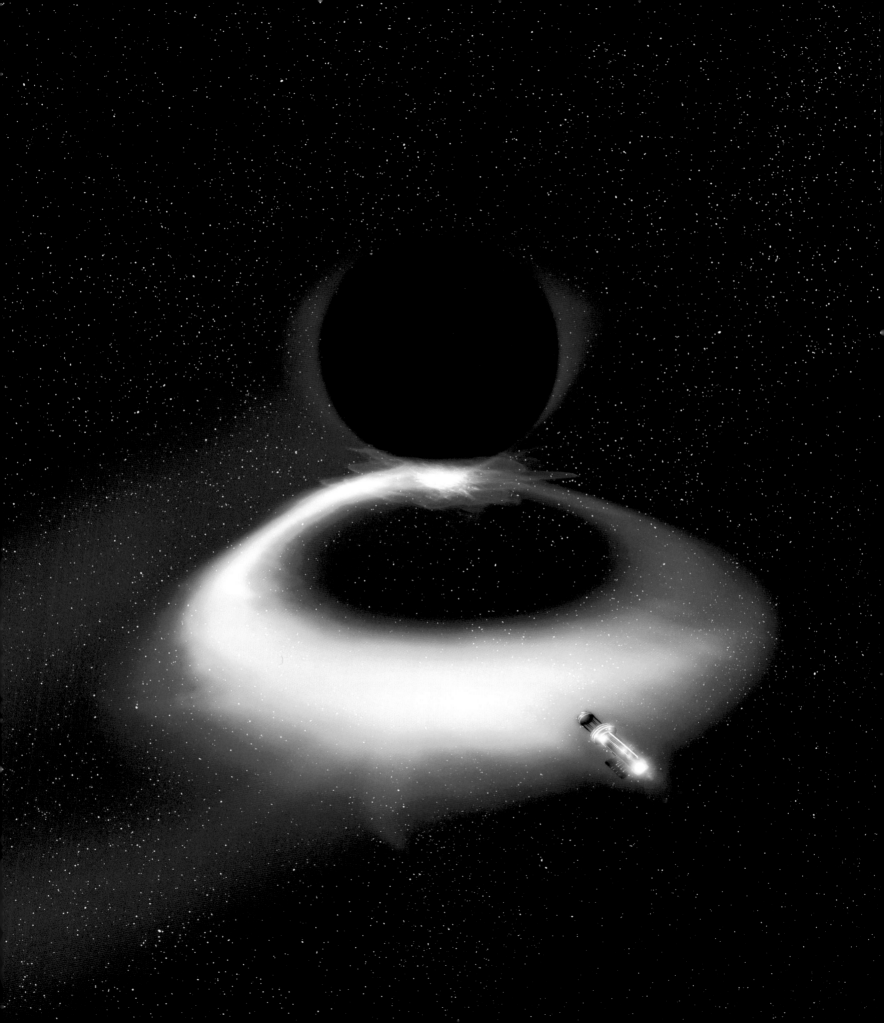

HAWKING RADIATION

In April 1975, Stephen Hawking published a remarkable paper on Communications in Mathematical Physics. The title of the paper was 'Particle Creation by Black Holes', a title that, as is often the case with the most historic of scientific publications, hid the profoundness of its content behind the most perfunctory of titles. In 21 pages Hawking argued that black holes are not completely black, that instead of the event horizon being a boundary beyond which nothing can escape they actually faintly glow, they have a temperature, they radiate. Hawking provided a theoretical proof of the existence of this Hawking Radiation, as it came to be known, in his now-famous equation.

$$T_H = \frac{\hbar c^3}{8\pi G k_B M}$$

With this equation, Hawking proved that black holes not only have a temperature but by emitting heat, by radiating, they must also be emitting mass. (Einstein's Theory of Special Relativity and equivalence principle, expressed in the equation $E=mc^2$, shows us that energy and mass are proportional, therefore if something is losing heat it must also be losing mass.) The mechanism by which Hawking suggested this happens takes us deep into the mindboggling world of quantum physics and the existence of a class of transient subatomic particles known as 'virtual particles'. Emerging from quantum field theory, virtual particles come into existence even in empty space, due to the briefest of energy fluctuations. These virtual particles exhibit some of the characteristics of ordinary particles, briefly coming into existence in pairs – a particle and an anti-particle that pop into the universe before almost immediately annihilating each other and disappearing again.

Hawking's great breakthrough was to demonstrate that when such pairs of particles formed around the event horizon of a black hole, one of them could fall into the black hole while the other could escape. In effect, this made it appear as if the black hole had radiated a particle away, reducing its energy and therefore mass while the energy and mass of the surrounding universe would have increased. The consequence of this discovery was profoundly revealing about the nature of a black hole, because if it was losing mass it meant that across unimaginable timescales black holes are slowly evaporating. Hawking had provided the theoretical basis for the death of every black hole in the Universe.

Over billions upon billions of years, time frames far longer than the current age of the Universe, every black hole including Sagittarius A* will gradually fade away. Getting smaller and smaller until 1,000,000,000,000,000,000,000,000,000,000,000,00 0,000,000,000,000,000,000,000,000,000,000,000,000,000,000,000,000,000,000, 000 years from now, in a final burst of light, Sagittarius A* will die.

Black holes, it seems, are no different from any other entity in the Universe. Another life story played out across the farthest reaches of time. Part of a universe with a beginning and an end, one where there was a first star and there will be a last, where life has begun and will inevitably end, and where even black holes, the most powerful objects in the Universe, will ultimately come to an end. It all seems to fit perfectly into the grand life story of our universe, a story that we have, with a fleeting collection of atoms, been able to piece together. But for all of its simplicity, it's a story that doesn't quite fit, because the fact that a black hole can die has raised one of the most profound conundrums in the history of physics – the information paradox.

HAWKING RADIATION
Particle–antiparticle pairs form from the quantum vacuum, with one particle falling into the black hole and the other escaping by quantum tunnelling. This mechanism allows small black holes to rapidly evaporate.

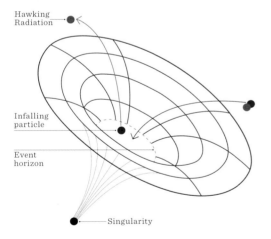

Hawking Radiation

Infalling particle

Event horizon

Singularity

Opposite: Artist's impression of a future probe approaching a black hole to capture Hawking Radiation, indicated as a blue glow.

Hawking Radiation

In 1975, Stephen Hawking published an amazing paper showing that black holes aren't absolutely completely black. They glow very, very faintly. They have a temperature associated with them. And you can write that temperature very simply in this beautiful equation, which ties together so many disparate areas of physics. You've got gravity in here, the mass of the black hole. You've got the speed of light. You've got something about the quantum world, subatomic physics. You've even got something about geometry. Tie all of these together and you get the temperature of a black hole.

If you have a temperature, you're glowing, you're radiating. When you put your hand next to something hot, you can feel it. And so over time, that radiation means that a black hole like Sagittarius A* is evaporating on timescales longer than the age of the Universe. This evaporation is going to mean that Sagittarius A* disappears to nothing.

So what happens to all the information that is contained in every bit of material that ever fell into a black hole? Not even light can escape from a black hole, so did we really lose all that information? And as a black hole evaporates, what happens to everything that is somehow encoded in all the Hawking Radiation? If we could sweep up all the Hawking Radiation, could we somehow reconstruct that history of everything that ever fell into the black hole?

In the very simplest terms, the information paradox emerges from a fundamental truth about our understanding of the Universe – the idea that nothing in the Universe is ever truly destroyed. Rip up a page of this book and it can be stuck back together; even go as far as burning this page and, despite its seeming destruction, nothing is actually destroyed. Every atom that made up the page before its destruction still exists, and could still, in principle, be painstakingly brought back together to recreate not just the page but every word on it, every piece of information reassembled. But if, as Hawking suggested, the atoms of this page ended up on the other side of an event horizon, lost in the abyss of a black hole like Sagittarius A*, never to return, where would all that information go to when our black hole finally disappears? If nothing can ultimately be destroyed, what happens to all the information when the holding pen it is being kept in has gone? Does the black hole return the information about everything that ever fell in back to the Universe, and if so, how? That is the essence of the information paradox, a conundrum that physicists have now fought over for almost 50 years. Even in death, the enigma that is a black hole mystifies us.

Up until very recently, our best theories have told us that everything that fell into a black hole, every sun, asteroid and planet is lost for eternity. We thought nothing, not even information, could escape from a black hole, but we've recently discovered our best-held theories are wrong. A growing number of physicists now believe that as Sagittarius A* evaporates away, the story of every meal it has ever eaten leaks

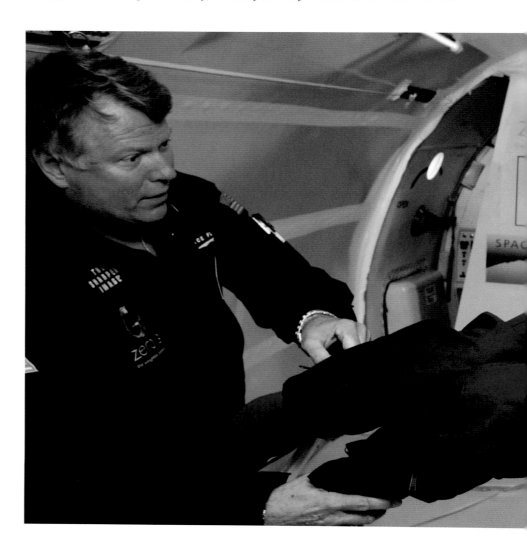

Right: Stephen Hawking in freefall flight. A modified Boeing 727 jet makes a set of steep dives that create short periods of weightlessness.

'Black holes are not the eternal prisons they were once thought. So if you feel you are in a black hole, don't give up. Black holes aren't tombs … they're gateways …'
Stephen Hawking

back out into the cosmos. The memory of all those worlds that fell into Sagittarius A* over the entire history of the Milky Way will still be there right at the very end. Fall into Sagittarius A*, and while you won't survive the journey, all the information that makes you unique, the raw ingredients that contain everything that makes you who you are, will endure – passing through the eye of the storm. A one-way trip to the end of the Universe. That is a tremendous achievement in itself; it took decades, but the real treasure lies in the explanation of how the information gets out.

Now, as much as this sounds like bizarre science-fiction, and nobody understands it fully, when the black hole has evaporated, it has gone through about half its life, and the interior of the black hole becomes in some sense the same place as the far-distant Hawking Radiation that was emitted eons before. It's as if spacetime wormholes open up between the interior of the black hole and the far reaches of the Universe, and that allows you to read the information inside.

However, there's a bigger picture here. If black holes connect places separated by trillions of light years of space – and eons of time – concepts so foundational to how we experience reality, they are not the immutable, fundamental features of the cosmos they might appear. We're a long way from comprehending all the secrets hidden inside our galaxy's supermassive black hole, but we are beginning to lift the veil. Far from being mere cosmic aberrations, supermassive black holes such as Sagittarius A* may hold the key to unlocking the Universe's deepest secrets.

ORIGINS

'The most terrifying fact about the Universe
is not that it is hostile but that it is indifferent.'
Stanley Kubrick

BEFORE THE DAWN

For most of its history the Universe has been all but empty, an infinite ocean stretching across an infinite expanse of time, but that is not the Universe that we see from our vantage point on Earth today. For us, our view is of a universe filled with islands of light, of galaxies – hundreds of billions of them, scattered in every direction – and each of these galaxies is home to hundreds of billions of stars, around which an even greater number of planets, countless other worlds, each unimaginably strange, hide in the darkness. And somewhere, drifting in this vast and fantastically varied cosmos, are you and me – miraculously improbable specks of stardust that can think and feel and wonder. We are the only atoms in the Universe that we know of that can ask questions about the Universe – what it is, why it appears as it does and

'In the beginning God created the Heavens and the Earth. The Earth was without form and void. And darkness was upon the face of the deep. Then God said, "Let there be light."'
Genesis

– the most profound question of all – 'How did all this come to be?' This is a question that has defined much of human history, the great conundrum about how our creation came about, but it's only in the last century or so that we've had the intellectual and technical tools to interrogate nature directly in our search for an answer. What we have found is a creation story unlike any other. Written in the language of mathematics and illustrated with the evidence above our heads, we have discovered a first moment in time, a beginning to the Universe 13.8 billion years ago, a moment we call the Big Bang. But that's not the ultimate story, not quite, because as we've explored deeper we've begun to suspect that there's more to it than that, and so we have embarked upon a search for evidence of the Universe – before the Big Bang.

A TIMELINE OF THE UNIVERSE

The Universe's evolution from the Big Bang to the present day. As the Universe cooled immediately after the Big Bang and the subsequent inflation, atomic nuclei began to form. After 300,000 years these nuclei combined with electrons to form atoms (mainly hydrogen). Within a million years, dense regions of atoms attracted nearby matter, forming gas clouds and, after about 400 million years, stars and then small galaxies. Gradually the galaxy formation rate decreased and the small galaxies merged into larger galaxies. Our solar system came into existence after about 9 billion years.

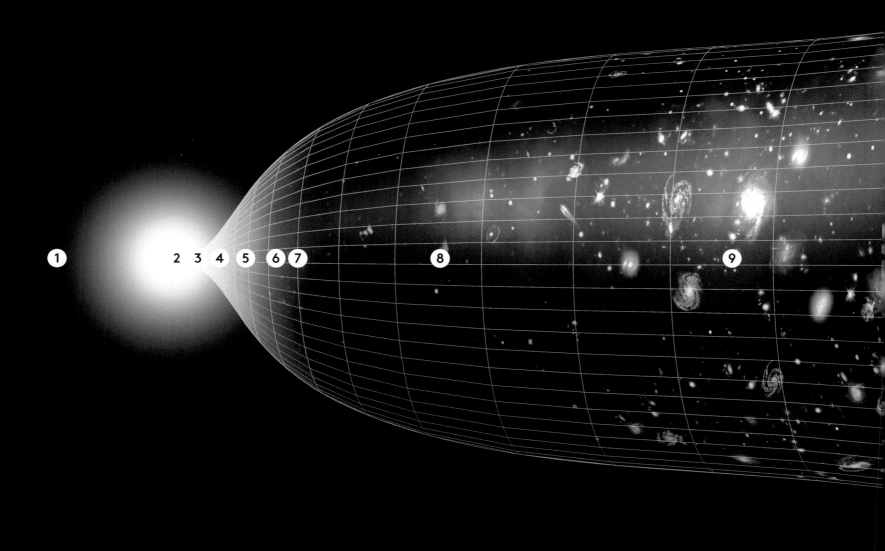

0
The Big Bang
The Universe begins 13.8 billion years ago with the Big Bang, creating both space and time.

From 10^{-32} seconds
Cosmic inflation smooths out irregularities in the early Universe and creates quantum fluctuations that will develop into structures.

From 10^{-30} seconds
The first particles **in the Universe** – protons and neutrons – start forming from quarks.

From 100 seconds
Ordinary matter particles are coupled to light, and dark matter particles start building structures.

From 300,000 years
The very first elements begin to form, mainly helium and hydrogen.

From 380,000 years
Ordinary matter particles decouple from light and the Cosmic Microwave Background is released.

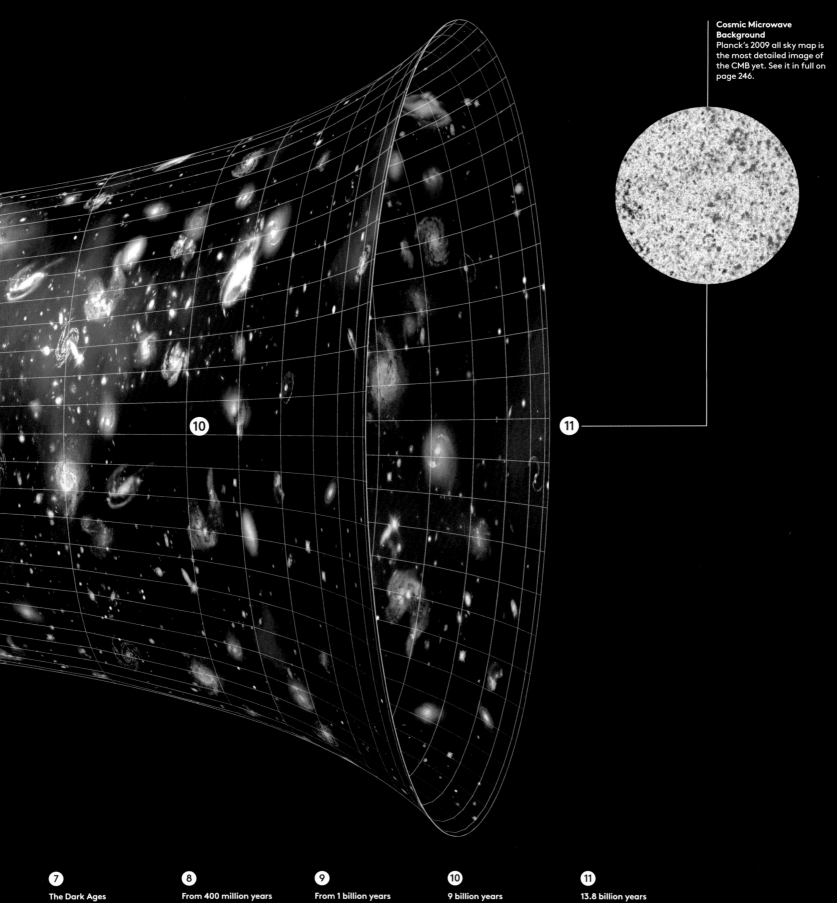

Cosmic Microwave Background
Planck's 2009 all sky map is the most detailed image of the CMB yet. See it in full on page 246.

(7)

The Dark Ages
Ordinary matter particles fall into the cosmic web structures created by dark matter.

(8)

From 400 million years
The first stars and the quasars begin to form. Subsequently, galaxies start forming.

(9)

From 1 billion years
The Universe expands; existing galaxies evolve into clusters and the larger superclusters.

(10)

9 billion years
The Sun and all the elements of our solar system form from the debris left behind by earlier stars.

(11)

13.8 billion years
Scientists discover Cosmic Background Radiation and begin to understand the timeline of our universe.

TIME MACHINES

Every time we stand under a clear sky and fix our gaze towards a star, we are making a deep and profound connection with the Universe. The light from that star is generated by the fusion of atoms deep within its core, before it is carried across millions upon millions of miles of space, then forms an unbroken stream of photons that enter your eye, bending through the lens before delivering their ancient energy to the light-sensitive receptors that connect you directly, physically, to that distant star.

Just think about that for a second: every star you have ever looked at connects with you directly across vast distances and immense eons of time. But travelling at 186,000 miles a second means that any starlight we look at will have taken a considerable amount of time to reach us. Light may travel fast, but on the scale of the Universe it is still relatively slow-moving. Take our own star as an example: 150 million kilometres is a minute distance on universal scales, and yet it still takes eight minutes for the light to journey from the Sun to the Earth. Which means that if the Sun was to suddenly disappear it would take eight minutes before we'd notice a thing. The vast distances of the heavens mean that there is not a star in the sky that isn't drawing us back, revealing its history to us, and with it the history of the Universe – a story told in light.

The next star out from the Earth is Proxima Centauri, a red dwarf star that is part of the triple Alpha Centauri star system. Currently closer to us than any other star, at a distance of 40,208,000,000,000 kilometres, it takes four years for the light of Proxima Centauri to make the short hop across this part of our galactic neighbourhood. Look up into the night sky towards Proxima and you are observing this tiny dim star as it was four years in the past.

This, however, is just the beginning. Our galaxy, the Milky Way, is around 100,000 light years across and contains perhaps as many as 400 billion stars, and that means as we look farther out we can travel further and further back in time. Kepler-444 is part of another triple-star system in the constellation Lyra, one of the oldest star systems in the Milky Way, and around this star is a collection of ancient planets. Situated 119 light years from Earth, the light we see from this star connects us back to the turn of the twentieth century, to a time when our feet were still firmly stuck on planet Earth and our view of the Universe had not yet stepped out from the confines of our own galaxy.

Moving onwards and outwards, from our position here on Earth there are around 50 million stars in our galaxy that sit around 2,000 light years away from us, stars like Deneb, in the constellation of Cygnus. One of the brightest lights in the Northern Hemisphere, this star has now exhausted its supplies of hydrogen and is on its way to becoming a red supergiant. Growing in brightness, it connects us back to biblical times as its ancient light ends its journey here on Earth.

All of these stars are visible with the naked eye, but the eye alone can only let us peer back so far. Of the billions of stars in our galaxy, fewer than 9,000 are visible to us unaided. The farthest of these lies in the constellation of Cassiopeia, a little over 4,000 light years away; it is 100,000 times more luminous than the Sun but appears as the faintest of stars, its light having left 4,000 years ago, while the Bronze Age was in full swing here on Earth, and Stonehenge was being completed.

Beyond this, if you've ever been lucky enough to witness the smudge of light that is the Andromeda Galaxy, you are looking at the light from a trillion suns that has been travelling for over 2.5 million years before it enters your eye. Light that has its origin in a time before even the earliest of human species had set foot on the Earth.

Above and opposite: Hubble Space Telescope in its orbit around the Earth.

Opposite top: Observations of the Moon from Sidereus Nuncius (Sidereal Messenger) by Galileo Galilei, 1610.

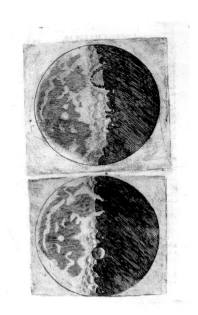

At the farthest reaches of our galaxy, we have glimpsed the light of a star called UDF 2457, one of the most distant objects we have seen within the main body of the Milky Way. Here, we are looking back almost 60,000 years, long before the beginning of our civilisation, to a time when *Homo sapiens* was just one human species vying for supremacy on Earth.

But to see further back than this, we cannot rely on the human eye alone – there are limits to its ability to capture light, focus it and detect the ancient photons that have made their way across the Universe. To be able to peer further into the darkness we need to work with technology, to develop machines that capture the faintest streams of ancient light, using lenses or mirrors that concentrate it and thus bring the past into focus. Since Galileo first used a telescope to explore the night sky in 1609, we have been on a continuous journey to build more and more powerful telescopes, machines that allow us to peer further out into the darkness, way beyond the limits of the human eye. These are in every sense of the word time machines, allowing us to glimpse back billions of years. The Universe is so vast, its lights so distant, we can not only attempt to understand its history, we can actually see it play out in the night sky and witness events from the deep past that have formed and shaped the Universe that we exist in today.

One such telescope has looked further back in time than any other; it is a time machine that has revealed the Universe's history to us in extraordinary detail. Launched into low Earth orbit by the Space Shuttle Discovery in 1990, the Hubble Space Telescope is still operational 30 years later and has allowed us to travel billions of years back in time. Hubble has given us unprecedented power to see the Universe's most ancient light from the distant shores of our own galaxy 60,000 years in the past.

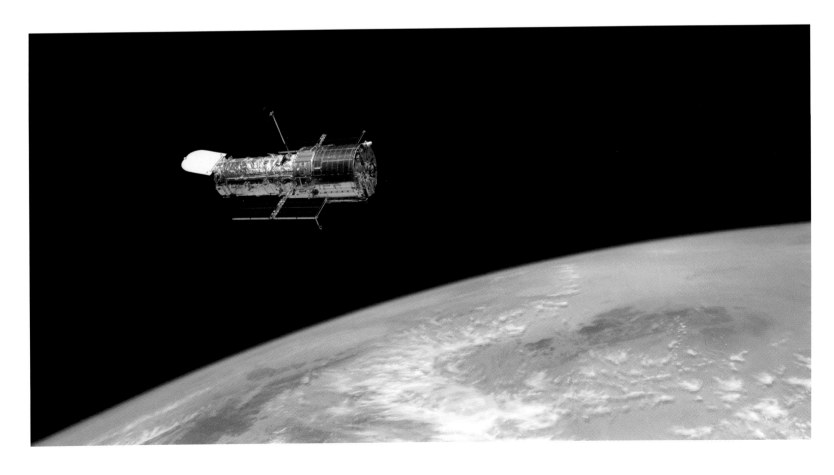

Below: As an ultra-dense white dwarf passes a small red star, its gravity is so great that it bends and magnifies the light of the red star, similar to the effect of using a lens.

Opposite: An 'Einstein Ring' occurs when light from two distant galaxies is bent by gravitational lensing from a massive elliptical galaxy in front of them and focused into a visible ring-like structure.

In April 2018, Hubble gave us a glimpse of the oldest, most distant individual star we have ever seen. Nicknamed Icarus (official name MACS J1149 Lensed Star 1), this fiercely hot, blue giant star sits 100 times further away from its closest individual star, in a distant, ancient, spiral galaxy. It's so far away that we would normally expect to be able to glimpse whole galaxies at this distance, but through a quirk of astronomical alignment we have been able to see a star as it was in the adolescent Universe just 4 billion years or so after the Big Bang. It has taken almost 10 billion years for the light from this ancient star to travel across the cosmos and into the eye of Hubble's waiting mirror, and the reason we can see it is because the light from this star was magnified not just by Hubble's powerful optics but by an effect called gravitational lensing.

First predicted by Einstein in his 1915 Theory of General Relativity, gravitational lensing is a phenomenon that occurs when light travelling towards us from a distant object passes by an extremely massive object, such as a galaxy cluster. In the case of Icarus, it not only required the chance alignment of the galaxy cluster MACS J1149 wandering between us and the distant star, but also the alignment of a massive object like a supernova within the galaxy cluster to create a process called microlensing, which magnifies the distant star thousands of times, allowing it to be glimpsed for a fleeting moment.

But even this is nowhere near the furthest Hubble has reached back into the past. In the summer of 2009, the Space Shuttle Atlantis undertook the fifth and final Hubble service mission, when it carried new upgrades to the telescope, including the Cosmic Origins Spectrograph (COS) and the Wide Field Camera 3, the most

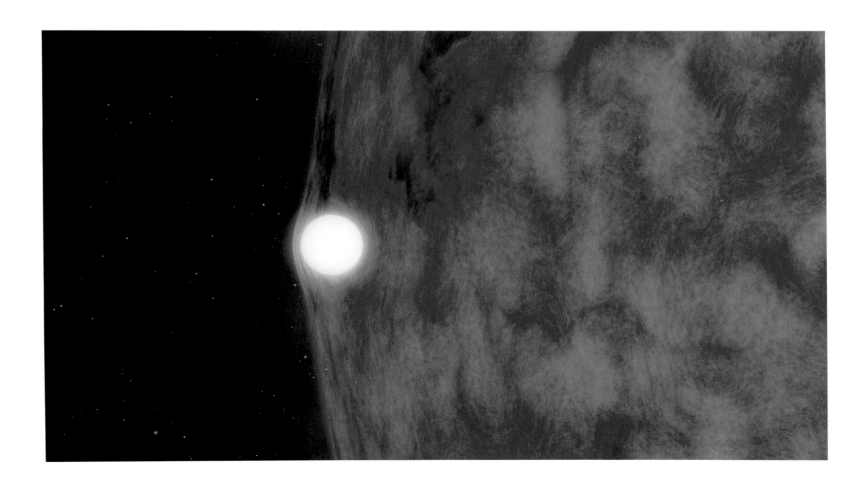

An insight into our Cosmic Dawn

The Hubble Space Telescope is the first great observatory and one of the greatest scientific missions in all of human history. Ultimately, all it is is a 2.4-metre piece of glass originally supposed to be part of a spy satellite from the Cold War. But we've turned it on our universe and it has enabled absolutely untold advances in our understanding of how our universe works. It has been a time machine. It has discovered other worlds. It has discovered how galaxies form and evolve. By putting that piece of glass above our atmosphere and putting really pristine optics and really advanced instrumentation at its focus, we have enabled more progress in the past 30 years than, frankly, we have in the past 30,000 years of astronomical history.

Never forget that the very act of looking into space means that you're looking back into the past. Time travel is not science fiction; you do it literally all the time. You have never witnessed the present in your entire life. So, if you're looking at someone roughly 1 foot away, you are looking at that person or thing or object as it was approximately one nanosecond or one-billionth of a second in the past. And if you step further away from it, 2 feet now, it's 2 nanoseconds in the past. And if you step 8 light years away from it, it is 8 years in the past and 230 million light years, a billion light years – and so on. With Hubble we can look at very, very, very distant objects and the very earliest stages of our universe; we can see our Cosmic Dawn.

Dr Grant Tremblay, Astrophysicist, Harvard Smithsonian Center

advanced optical detector ever used in space. These two new instruments were installed during the mission's five spacewalks and other repairs were also undertaken, to bring the Hubble Space Telescope to the peak of its capabilities. So, with its upgrade in place, Hubble began the largest project in its history. The Cosmic Assembly Near-Infrared Deep Extragalactic Legacy Survey, otherwise known as CANDELS, took place between 2010 and 2013 using over 60 continuous days of Hubble's observing time across 902 assigned orbits. As the telescope stared into an exact patch of apparent darkness day after day, it began to collect the most distant light ever seen. The resulting image is at first sight not the most impressive nor the most beautiful of the thousands of images that Hubble has captured over its lifetime, but when you realise what is contained within it, it soon becomes one of the most profound pictures we have ever seen. This is Hubble's Ultra Deep Field, an image that is not filled with individual stars but individual galaxies stretching across half the observable Universe.

The oldest and most distant galaxies appear red here, because the light from them has been travelling for so long that the expansion of the Universe has caused it to be stretched, or 'redshifted' to longer wavelengths. When this phenomenon occurs, every distant object in the Universe appears to be receding from us because its light is stretched to longer, redder wavelengths as it travels to our telescopes through expanding space. The greater the redshift, the farther away the galaxy, and these galaxies are so distant they are allowing us to peer back to a time less than 1 billion years after the Big Bang. It was among these that we discovered the most ancient galaxy we have ever recorded.

THE HUBBLE SPACE TELESCOPE

Communications system | Equipment bay | High gain antenna | Crew handrails | Aperture door

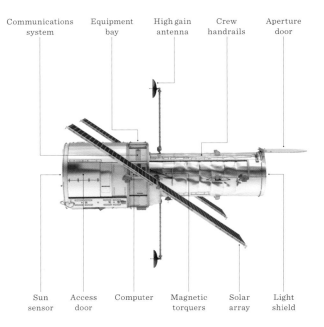

Sun sensor | Access door | Computer | Magnetic torquers | Solar array | Light shield

Below: Westerlund 2, a cluster of about 3,000 stars, resides in the raucous stellar breeding ground Gum 29, 20,000 light years from Earth.

Right: Backlit wisps along the Horsehead Nebula's upper ridge are illuminated by Sigma Orionis, a young five-star system.

'The Universe is like a beautiful abstract painting; we have to look at every single part of it to understand it, and everybody sees it differently.'
Rana Ezzeddine, Astronomer, University of Florida

Opposite: The Hubble Ultra Deep Field includes 5,500 galaxies, some of them among the most distant ever identified.

Above: Hubble captures the Sombrero Galaxy, a spiral galaxy around 50 million light years away, in the constellation Virgo.

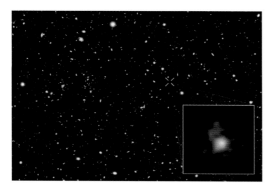

Searching for answers in the Dark Ages

The formation of the first stars and when that actually took place is a question that is under debate and is one of considerable interest at the moment. There was certainly a very extended period in our universe, which is sometimes known as the Cosmic Dark Ages, where there was no light in the cosmos and there was only a certain point relatively late on in cosmic evolution where the first stars came into being and started to light up the cosmos for the very first time.

Even though the name the 'Cosmic Dark Ages' suggests that there might not have been anything particularly interesting going on, it actually turns out that the bedrock for everything that then transpired in the Universe to make it look like how it does today was being sculpted at that time. So the Dark Ages was a moment where dark matter was really kind of laying the groundwork for the construction of what's known as the cosmic web.

So, these are the kinds of questions that we as astronomers are asking ourselves – how can we actually try to measure and detect the very first stars and galaxies in the Universe to help us understand when the first dark matter structures were created and what the nature of that dark matter might be?
Sownak Bose, Researcher,
Center for Astrophysics, Harvard

Announced in March 2016, the discovery of GN-z11 shattered all previous cosmic distance records as the farthest galaxy ever seen in the Universe, but it was also the oldest object ever seen. At 13.4 billion light years away, it is a galaxy that formed just 400 million years after the Big Bang, at a time when the Universe itself was still taking shape and was just 3 per cent of its current age. This is light from some of the first stars, light from the dawn of creation. We know this because we've been able to measure precisely how much this ancient light has been redshifted, which gives us a precise indication of its distance and age. Before GN-z11, the most distant galaxy measured spectroscopically had a redshift of 8.68, equating to a distance of 13.2 billion light years, but using Hubble we've been able to confirm GN-z11 to be at a redshift of 11.1, meaning we are looking back nearly 200 million years closer to the Big Bang.

At just a quarter of the size of our galactic home and containing just 1 per cent of its mass – tiny by the standards of the Milky Way – GN-z11 was a strange galaxy, but what it lacked in size it made up for in activity. Bursting with enormous and incredibly violent stars, developing at a rate 20 times faster than our own galaxy, we think it was filled with hot blue stars that were no more than 40 million years old. We did not expect to find such a massive galaxy existing so soon after the very first stars had started to form, and this one was forming fast and burning bright.

The fact we have glimpsed so far back in time is almost incomprehensible, but armed with this knowledge we can try to travel even further into this ancient galaxy. Applying what we now know to these young stars, we can also imagine that there were worlds hiding in this light, some of the first planets to exist in the history of the Universe. These would have been strange, primordial worlds, ravaged by the intense radiation generated by their proximity to the violent stars, and it was perhaps on the surface of one of these planets that something spectacular occurred – that dawn broke in the Universe for the very first time.

We have no idea when or where the first planets formed around the first stars. GN-z11 might have borne witness to one of these first dawns, but looking back so far in time, it's impossible to know with any certainty. What we do know is that this dawn was not the first moment in the story of the Universe; the stars and planets had to come from somewhere, from a time that we know was hidden in the dark; a time that not even Hubble can glimpse for us because there were no galaxies before GN-z11, no planets, not even stars. In our journey back through the history of the Universe, we reach a point where it seems we can see no further. Astronomers call this long dark night the Cosmic Dark Ages. And it's here – in the profound gloom of the Cosmic Dark Ages, in a universe before stars of any kind – that the secret to the Universe's origin must lie.

Even the most powerful telescopes imaginable will never penetrate the gloom of the Cosmic Dark Ages, so to continue our journey back towards the origin of the Universe, we must turn our gaze in search of other clues that can lead us to the ultimate origin of you, me and everything else.

Stuck in our small corner of the galaxy, we've been able to travel to the farthest reaches of the Universe by following the light backwards through time, but in the last 100 years starlight has been used to guide us through its history in other ways as well. Travelling across the Universe for millions if not billions of years, it is imprinted with information about the way it has evolved, a secret history that we are learning to read among the stars.

HISTORY IN MOTION

'The magnitude of this velocity, which is the greatest hitherto observed, raises the question of whether the velocity-like displacement might not be due to some other cause, but I believe we have at present no other interpretation for it.'
Vesto Slipher

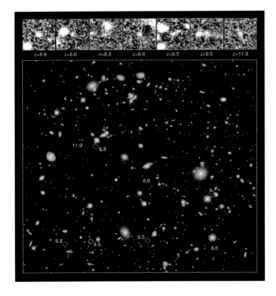

At the beginning of the twentieth century, our view of the Universe remained confined by space and time. We had yet to see beyond the shores of our own galaxy, and for many of the great astronomers of the day, the simplest conclusion was that the Universe did not extend beyond the Milky Way, beyond this one vast island of light.

We had also yet to get any real understanding of the past or future of the Universe, no sense of where it had come from and how it would evolve. At this time, the consensus position was that this was an eternal universe, one without a life story. Many of the greatest minds in science, including Albert Einstein, believed the cosmos stretched infinitely forwards and backwards in time. But by 1912, the first clues to a deeper truth were beginning to emerge.

While working at the Lowell Observatory in Flagstaff, Arizona, Vesto Slipher, an American astronomer, began measuring the colour of light, the spectral lines of distant galaxies or nebulae, as they were then known. Looking first at the Andromeda Nebula, Slipher noticed something particular about the light coming from this distant body – it was redshifted. Slipher concluded this could mean only one thing, that the nebula was speeding away from us at exceptional speed, a velocity he estimated to be around 300 kilometres per second. Following up on this finding, Slipher used the same technique to observe the light of 15 other nebulae and found that almost all of them were also speeding away from us. It seemed that everywhere we looked, the Universe was rushing away at incredible speed. It was an observation with profound significance, but for Slipher it was an observation that came a little too soon. At the time of his discovery we had yet to understand that these nebulae were in fact galaxies, vast islands of billions of stars. The deep cosmological implications of his finding would remain hidden for another 15 years.

It would take the work of the most famous astronomer of the twenty-first century to bring Slipher's findings back into the light. Working at the Mount Wilson Observatory in California, Edwin Hubble had to hand the 2.5-metre Hooker Telescope, the most powerful of its kind on Earth at that time. Arriving at the observatory in 1919, Hubble used the telescope to observe a type of star known as Cepheid variables. These stars had a precious characteristic that had been discovered a few years earlier by American astronomer Henrietta Leavitt at the Harvard College Observatory. After studying the changing luminosity of hundreds of these variables, Leavitt had shown that the intrinsic brightness of these stars could be calculated and so used as a measure of distance – what is known in astronomy as a standard candle, a yardstick for the Universe. Put simply, the farther away an object is, the dimmer it appears, so if you know the brightness of each and every one of these stars you can measure their distance from Earth.

This was a massive breakthrough and it allowed Hubble to build on Leavitt's work so that a few years later he was able to discover a number of Cepheid variables in the Andromeda Nebula. When he measured the distance to these stars, it was clear that it was of such magnitude that the Cepheid variables could not be part of our own galaxy, which allowed Hubble in 1924 to take us out of our island universe, revealing nebulae like Andromeda to be whole other galaxies, islands of stars beyond our own.

Opposite: Lowell Observatory, founded in 1894, is one of the oldest observatories in the United States, and it's where the dwarf planet Pluto was discovered in 1930.

Above: Hubble Ultra Deep Field 2012, pinpointing a population of galaxies at redshifts between 9 and 12, the most distant ever seen.

PLOTTING HUBBLE'S LAW

When the velocity of galaxies (y axis) is plotted against the distance to the galaxies (x axis), the resulting graph means the two values are proportional, related by the Hubble Constant. This discovery by Edwin Hubble supported the theory that the Universe is expanding and enabled scientists to calculate its age.

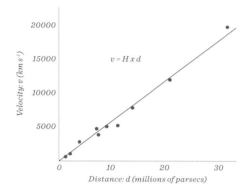

Using the Cepheid variables, Hubble now had the means to measure the distance of dozens of other galaxies, galaxies that he knew from Slipher's redshift results a decade before were also speeding away from us. Leaning on the work of many others, including Slipher, Belgian astronomer Georges Lemaître and Milton Humason (his assistant at Mount Wilson), Hubble measured the distance to 24 galaxies, then correlated these measurements with their velocity as indicated by the redshift of their light. The resulting plot is captured in the now-famous graph (left), which revealed a simple but profound truth and in an instant changed our understanding of the Universe forever. What Hubble observed for the first time was a simple correlation between the distance and speed of each of these galaxies. Each was moving away from us at a speed proportional to their distance from Earth – a relationship that became known as Hubble's Law (or more recently The Hubble–Lemaître Law, but more on that later). The farther away the galaxy, the faster it seemed to be moving. Hubble had made an observation with one single, profound consequence: this was not a static, eternal universe, this was a universe that seemed to be expanding. Speaking at the annual meeting of the National Academy of Sciences in 1934, Hubble laid out the potential consequence of his discovery to the world, stating:

Left: Edwin Hubble changed our understanding of the Universe when he proved that it is not static, but expanding rapidly.

Stretching space through time

When physicists talk about the expanding Universe,
what we really mean is that space itself is stretching
between where we're sitting here in the Universe
and where distant objects far away from us are. So the
space in between is actually stretching, almost like
a kind of trampoline or a rubber sheet. It's actually
growing over time. And what that means is that
objects are getting carried further away from us and
in fact from every other object – it's a remarkably
uniform stretching of this kind of fabric of space in
every direction over time.

It can feel pretty unsettling to think that this sort
of seemingly solid substance, this kind of scaffolding
around which we move all the time might actually
not be so solid, that space itself is wobbly. It can bend
and warp and it can stretch. It can get bent out of
shape near enormous, massive objects like the Sun
or a black hole, or even a whole galaxy. So its shape
can be deformed in one region of space and it actually
can stretch and get out and carry things further and
further apart over time. So what we think of as this
solid grid work within which we live our cosmic lives
is actually not so solid after all.
David Kaiser, Physicist, MIT

Above: Edwin Hubble,
photographed with the
Hale Telescope (the world's
largest until 1976) at Palomar
Observatory in California.

'The present distribution of nebulae can be represented on the assumption that they were once jammed together in one particular region of space and at a particular instant about 2 billion years ago they started rushing away in all directions at various velocities. The expanding Universe, with its momentary dimensions as previously described, is the latest widely accepted development in cosmology.'

Today, we take the concept of an expanding universe for granted, but it took many decades to turn Hubble's initial observations into the basis of the fundamental creation story as told by science today. The idea of an expanding universe ran counter to hundreds, if not thousands, of years of thought, impossible to align with the widely held belief in the concept of a steady-state universe – a universe that never grew older but instead was ageless, with new galaxies, stars, planets and perhaps even life, as part of an endless cycle of creation. It wouldn't be until the late 1960s and early 1970s that the alternative explanation would fully take hold, the idea that 'the geography of the Universe can be turned into a history', that the Universe has a beginning and an end.

The key challenge in turning Hubble's controversial observations into a definitive history of the Universe has taken decades upon decades of work as scientists have repeatedly attempted to increase the accuracy of Hubble's initial measurements by endlessly refining our ability to measure a single number at the heart of that initial discovery, known as the Hubble Constant.

The Hubble Constant is the unit of measurement used to describe the expansion of the Universe. It is most frequently expressed as the speed of a galaxy 1 megaparsec away (an astronomical unit that represents 3.09×10^{19} kilometres) and is the key number needed to not only express the expansion of the Universe, but by also winding back the expansion it reveals the age of the Universe, too. Back in 1929, when Hubble plotted his initial data and made the first calculations, he came up with a number of around 501 kilometres per second per megaparsec – a value that had one significant issue that was impossible to ignore. This number implied that the Universe was in fact younger than the Solar System – a cosmos no more than 2 billion years old, when we knew the Sun and Earth had existed for almost 5 billion years. This alone made Hubble doubt the validity of his own findings, suggesting that the redshift observed in distant nebulae might be down to an unknown property of space rather than a true measurement of their velocity.

Commenting on the conundrum himself in 1934, Hubble knew this uncertainty around his findings would only be reduced by the collection of more accurate data and that the ability to refine the measurement of his constant would come with the advancement of technology. This, Hubble hoped, would come from observations with the new telescope that was under construction at the California Institute of Technology at the time: 'Further radical advances in cosmology will probably await the accumulation of more observational data …' explained Hubble, '… and will definitely answer the question of the interpretation of redshifts, whether or not they represent actual motion; and if they do represent motions – if the Universe is expanding – may indicate the particular type of expansion. This prospect is the climax of the story.'

Below: Double star OGLE-LMC-CEP0227 in the Large Magellanic Cloud (LMC). The smaller of the two stars is a pulsating Cepheid variable.

Bottom: Gaia shows the stellar densities of the LMC and SMC. Red indicates giants, green main-sequence and blue, young stars.

Although Hubble could barely have imagined it at the time, the ultimate source of that data would come from a telescope not sitting here on Earth but orbiting the planet 540 kilometres above its surface. The telescope bearing the great astronomer's name has perhaps done more to change our view of the cosmos than any other single piece of technology in human history. Rid of the distorting effects of the Earth's atmosphere, the Hubble Space Telescope has allowed us to peer farther out into the cosmos with more power and precision than Hubble himself could have dreamed of. With its clear view of the heavens, it has allowed us to measure the distance to faint stars and galaxies with ever-greater accuracy over its 30-year lifetime. With this greater accuracy has come a more reliable set of data to calculate the Hubble Constant, which has been used as the basis for our estimates of a wide variety of fundamental cosmological parameters, including the age of the Universe.

Before the Hubble Space Telescope, the most accurate data then available only allowed us to calculate the age of the Universe as somewhere between 9.7 and 19.5 billion years. But Hubble's power changed all of that. Using the telescope, it was possible to make a precise measurement of the distance to a collection of Cepheid variable stars 56 million light years away in the Virgo Cluster of galaxies. With this new data, astronomers were able to refine the value of the Hubble Constant and therefore the age of the Universe to an error of only 10 per cent by the late 1990s. But that was just the beginning, because as the Hubble Space Telescope was updated in a series of service missions throughout the first 20 years of its operation, the precision with which the telescope could operate dramatically improved, allowing the calculation of the Hubble Constant to become ever more precise.

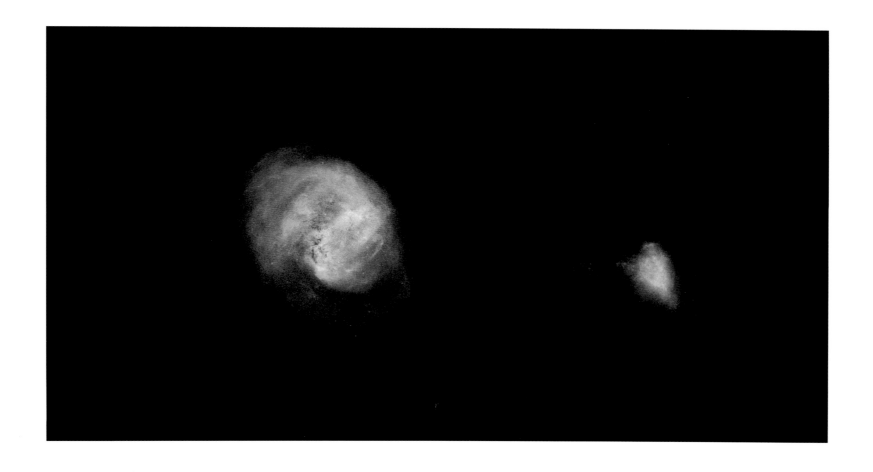

CALCULATING THE HUBBLE CONSTANT
Astronomers build a three-step 'cosmic distance ladder' to calculate the Hubble Constant, by measuring the distance of Cepheid variables nearby and farther afield and comparing both to galaxies with Type 1a supernovae and the redshifted light of distant galaxies.

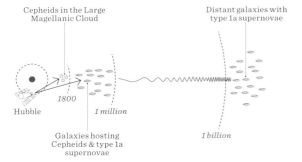

Cepheids in the Large
Magellanic Cloud

Distant galaxies with
type 1a supernovae

Hubble

1800

1 million

Galaxies hosting
Cepheids & type 1a
supernovae

1 billion

Calculating the Hubble Constant relies on being able to do two things at once: measuring the redshift of distant stars and galaxies through precision spectroscopy; and measuring the precise distance to those objects with extreme accuracy. To maximise the precision of these calculations, astronomers need to measure these two values in objects increasingly farther away from Earth; the farther away we can measure an object the more accurate our calculation of the Hubble Constant can be. To achieve this, astronomers have been continuously streamlining and strengthening the construction of a cosmic distance ladder in recent years. The ladder is built from a collection of techniques that when combined allow us to step out able to accurately measure stars and galaxies near and far.

Hubble has been instrumental in the construction of this 'ladder', providing the firepower to measure objects in the Milky Way, nearby galaxies and outwards into the cosmos. For local distances, Cepheid variables, the same type of star that Hubble used to first calculate the distance to our nearest galaxies, are still the most useful yardstick, the standard candle for astronomical measurements to the edge of our galaxy where Cepheid variables can be observed in the Large Magellanic Cloud. Even farther out, to the Andromeda Galaxy and beyond to galaxies within tens of millions of light years away, Cepheid variables remain a brilliant yardstick for measuring distance, but if you really want to increase the precision of the Hubble Constant you need to be able to measure distances beyond which these variable stars become just too dim to use. To measure the distance to the farthest galaxies, astronomers have turned to using a different standard candle – one that is only fleeting but when it does shine creates the briefest but most spectacular shows in the Universe, allowing us to trace our ladder even further outwards into the cosmos.

Right: The central Milky Way with the Trifid Nebula at centre right and two newly discovered Cepheid variable stars visible in infrared.

THE DANCE OF DEATH

'Herschel removed the speckled tent-roof from the world and exposed the immeasurable deeps of space, dim-flecked with fleets of colossal suns sailing their billion-leagued remoteness.'
Mark Twain

Above: William Herschel, pioneer explorer of the skies, discovered thousands of objects that we now know to be distant galaxies, including NGC 2525 in 1791.

Seventy million light years from Earth, NGC 2525 is a beautiful barred spiral galaxy, thought to be at least 60,000 light years in diameter and with a supermassive black hole lurking at its centre. The galaxy is part of the constellation of Puppis, in the Southern Hemisphere, and we've been looking at it since William Herschel first set eyes on it in February 1791. Little did we know that 230 years later it would provide the most spectacular of events that would allow us to use it as one of the most distant galaxies in our attempt to measure the expansion of the Universe and calculate its age more accurately than ever before.

Although we could never have known it at that time, one of the Universe's stranger beasts, a white dwarf, was hiding in the darkness of this distant spiral galaxy in a precariously fragile state of existence. White dwarfs are the remnants of stars, just like our sun. Not big enough to create a black hole or neutron star when they reach the end of their lives, instead these stellar remnants become incredibly dense, incredibly faint objects, often hidden by the light of their younger, brighter neighbours. Thought to be the final evolutionary stage of almost all main-sequence stars between 0.07 and 10 solar masses, there are estimated to be in the region of 10 billion white dwarfs in our galaxy alone. In the last stages of its life, a star on its way to becoming a white dwarf will have used up almost all of its supply of hydrogen, with a remaining core made up of mostly carbon and oxygen. As it expels most of its outer remaining material and creates a planetary nebula, it only leaves behind the extremely dense carbon-oxygen core, no bigger than the Earth but with a mass equivalent to that of our sun.

For most white dwarfs, the future holds only a long, slow journey to obscurity. Hot when they first form from the remnant heat of the now-dead star, white dwarfs no longer undergo fusion, so with no energy source to power them, this residual heat slowly fades. As the core cools, the matter within becomes crystallised, becoming in effect stellar diamonds, dimly shining in the darkness. The ultimate destiny for almost every white dwarf when it no longer emits any heat or light is what is known as a black dwarf, and yet despite the theoretical existence of these objects we don't believe any exist in the Universe yet. White dwarfs take so long to cool that we calculate there isn't one that's old enough to have reached its final stage. Even at 13.8 billion years old, the Universe is too young to have yet given birth to its first black dwarf.

Yet for a tiny number of white dwarfs the future is anything but dark. The immense density of these objects means that they generate an enormous gravitational pull, and that makes some of them ticking time-bombs. If the mass of a white dwarf is able to increase above a critical limit known as the Chandrasekhar Limit (equivalent to 1.4 solar masses), it is no longer stable and gravity will take over, crushing the white dwarf and diverting its destiny onto a very different path.

And that's exactly what happened to a white dwarf on the outer edges of the galaxy NGC 2525 at the very start of 2018. This white dwarf was not alone – for millions of years it had been locked in orbit around a red giant – but this was a dance that couldn't last forever. As they spun around each other, the white dwarf, with its powerful gravitational pull, began to steal gas and plasma from its vast neighbour. Feeding off the giant its mass increased, creeping slowly upwards until the mass of the dwarf star reached that critical limit, so dense that the very matter it was made of could resist the pressure no more. With the relentless pull of gravity taking over, the core could no longer hold itself up and in the briefest of moments it went from stability to collapse.

PUPPIS

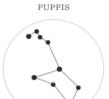

Seventy million light years away in the forested mountains outside Yamagata City, in Japan, one of the world's most prolific amateur astronomers, Koichi Itagaki, was watching. It was 15 January 2018 when Itagaki noticed what looked like a brightening star in the direction of the constellation Puppis. With over 80 discoveries to his name by this point, Itagaki had plenty of experience that told him exactly what he was looking at – the birth of a supernova in a galaxy far, far away – and this was a supernova that had yet to reach its peak magnitude, not yet fully self-detonated into oblivion. SN 2018gv, as it came to be known, had given us just enough warning for us to be able to witness it light up the night.

Entirely impossible to predict, the discovery of a new supernova is always of immediate interest to astronomers around the world, and SN 2018gv was no exception. With 10 to 15 days to go before it reached its maximum, fatal illumination, the team in charge of the Hubble Space Telescope swooped into action. With Itagaki acting as our cosmic lookout, the Hubble team were able to turn the all-powerful eye of the telescope towards his discovery and start observing it just a few days later. Watching and recording the radiance of an object 70 million light years away, it was so bright that the Hubble team were able to make a remarkable movie out of it as it became ever brighter.

Left: The death throes of a star expel layers of superheated gas that tear across space at more than 965,000 km/hour.

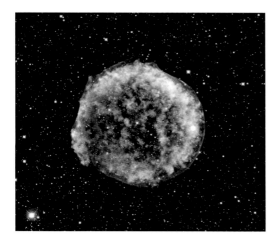

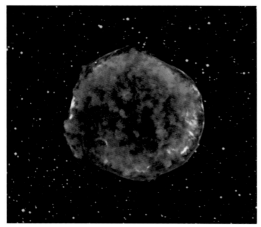

But this was not just an everyday film star, this was a very special type of star, a gift to astronomers attempting to measure the distance of galaxies far across the Universe. Because, although supernovae like these shine for just a few days, as they do so they cast a profound and powerful light out across the Universe.

Astronomers have given a name to this kind of supernova – Type 1a – and they really are a gift from nature. As they suck in mass from their neighbouring stars, white dwarfs destined to become 1a supernovae slowly reach a point where the pressure and density created by the additional mass raises the temperature of the dead star. Teetering on the edge, we think the star can stay in this simmering phase for around 1,000 years before it reaches the temperature needed for the reignition of the star's fusion reactions. In this moment, a star that burned bright for perhaps 8 or 10 billion years, that after its death lived in the shadows as a white dwarf for millions upon millions of years and then, after feeding on a neighbour, gradually brought warmth back into its core over a thousand years, comes back to life and reignites as the ignition temperature for carbon fusion is reached. And within just a few seconds of the star relighting, a runaway reaction begins; unable to regulate its temperature, this once-dead star rises from the ashes and transforms at enormous speed. Carbon and hydrogen fuse into heavier elements in the space of a few seconds, releasing huge amounts of energy that raise the internal temperature to billions of degrees in a matter of moments. With such a rapid release of energy, the star can no longer contain itself and the matter making up the white dwarf begins

Above: Tycho supernova remnant is a moving cloud of hot gas and dust formed after the Type 1a supernova of a white dwarf star in 1572.

Below: The density around a star 10 minutes after its companion goes supernova. Density moves from red (high) to dark blue on the left, and from black to red on the right. Ejected matter from the exploding star forms complex vortices, shock fronts and eddies as it interacts with this star.

to fly apart at unimaginable speeds. And this is what Koichi Itagaki was the first to witness here on Earth on that January day in 2018: a white dwarf that exploded 70 million years ago in the most violent of cosmic events, sending out a stellar shockwave travelling at up to 20,000 kilometres per second that creates a brief burning light 5 billion times brighter than the Sun.

As well as being incredibly violent, we now know that Type 1a supernovae are also incredibly predictable. With a characteristic light curve for each and every one that plots out their light curve over time, a group of Chilean and US astronomers working on a project known as the Calán/Tololo Supernova Survey were able to show that these supernovae, just like Leavitt's Cepheid variables discovered almost 100 years earlier, could be used as standard candles. All exploding in the same way and all shining with the same brightness, it means that if you see one that is dimmer it must be further away. And unlike the Cepheid variables, because they are so bright we can see them tens of billions of light years away, which means we can accurately measure the distance to galaxies right out to the edge of the observable Universe, all the way to galaxies like NGC 2525.

And that's exactly what we did, starting in February 2018. As this Type 1a supernova burned through its brief but bright life, the Hubble Space Telescope could use its light to measure the distance to that faint galaxy NGC 2525. Combining this with the redshift value from both the light of the supernova and the galaxy itself, SN 2018gv was able to give us another data point in the construction of the most

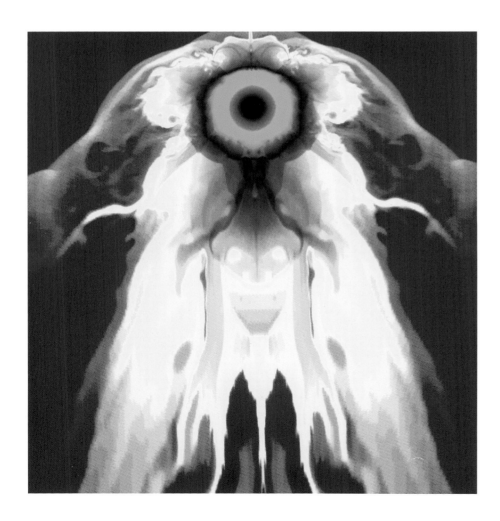

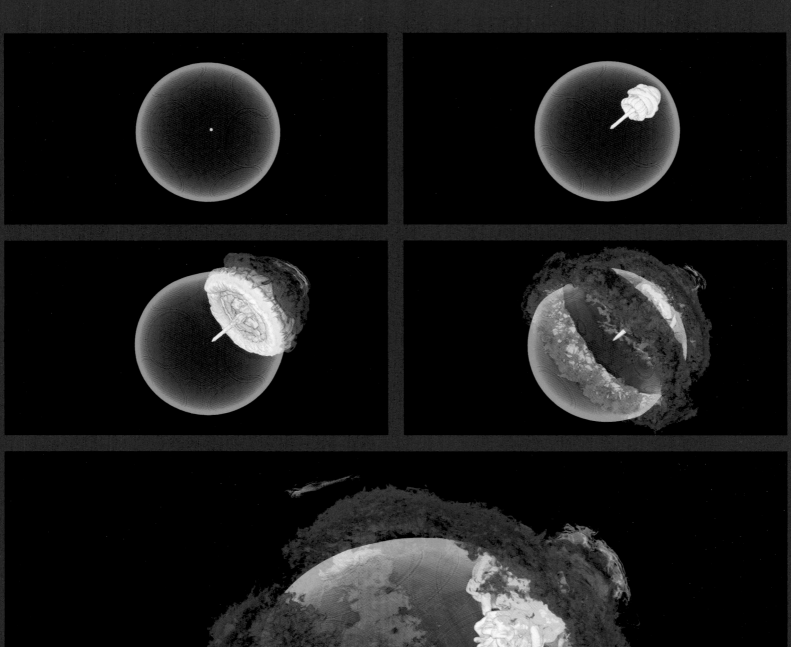
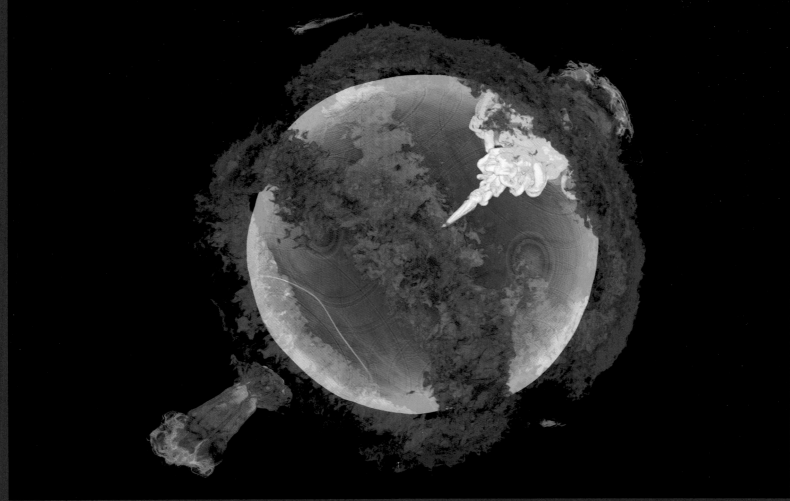

LIGHT CURVE
This plot of luminosity (relative to the Sun, L0) versus time shows
the characteristic light curve for a Type 1a supernova. The peak
is primarily due to the decay of nickel (Ni), while the later stage is
powered by cobalt (Co).

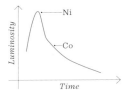

accurate cosmic ladder to date, allowing the measurement of the Hubble Constant
to improve to such an extent that from 10 per cent uncertainty at the start of the
2000s we now have a measurement of the expansion rate of the Universe (at the
time of writing, in May 2021) that has a degree of accuracy of 98.1 per cent.

This makes our new estimate of the Hubble Constant 74.03 +/- 1.42 kilometres
per second per megaparsec. Wind that number back and it gives us an estimated
age of the Universe to be 13.8xx +/- xx billion years since it all began.

Job done, it seems, but as is often the case in science, one data set is never
enough. Although the Hubble Space Telescope has enabled us to massively increase
the accuracy of our calculations of the Hubble Constant over the last 30 years,
the data it published in 2019 also created a new problem: the increased accuracy
of the Hubble data sharpened the discrepancy between its findings and those of
the other main method for calculating the Hubble Constant. In the last 20 years,
scientists have also been using a method that takes our images of the earliest light
in the Universe, the cosmic microwave background radiation (CMB), and uses it
to fast-forward through the life of the Universe to come up with another method
for measuring the Hubble Constant and therefore the age of the Universe. To start
with, measurements using both techniques were similar enough not to cause any
concern, but as Hubble increased the power of its eye, so did the other side. The
European Space Agency (ESA) Planck Telescope, launched in May 2009, has
enabled us to massively sharpen the data for the CMB approach in the same way
that Hubble has done for the cosmic ladder technique, and the result has been an
ever-increasing discrepancy between the two main methods for calculating the
Hubble Constant.

With the Planck team coming up with values as low as 67, we now have two vastly
different values for the expansion rate of the Universe, and two values for the age
of the Universe that differ by as much as 2 billion years – two values that can't both
be right. A problem, yes, but an exciting one at that, because no one can be sure
what is causing the discrepancy. Such situations are not rare in science, but usually
as the reliability of the data increases, the variation between two different methods
decreases until it disappears. In this case, however, the opposite has happened.
Whether it's an unknown flaw in either of the experimental methods, we do not
know, but when a gap like this appears in our knowledge there is also the most
exciting of possibilities: the prospect that both calculations are correct but that
the discrepancy suggests a fundamental gap in our current understanding of the
Universe. A gap that some scientists now think points in the direction of as yet
undiscovered new phenomena, the promise of new physics and a new understanding
of how the Universe works. And it doesn't get more exciting than that.

While astronomers and cosmologists continue to fight over the precise value of
the Hubble Constant, one thing that doesn't change is our quest to find the origins
of the Universe. If the Universe is expanding, it follows that everything was once
much closer together, and if we want to understand how it all began we have to go
back to a moment when our universe was much smaller and everything in it was
much closer together. We have to go back to a time before the Earth and our sun
existed, to a time before the first galaxies like GN-z11 came into existence, to a time
before the Cosmic Dark Ages. And if we keep travelling back beyond the darkness,
we end up at the most famous moment in the history of the Universe – the Big Bang.

Opposite: A computer model
of a white dwarf star the
size of the Earth but with the
mass of the Sun exploding
into a Type 1a Supernova.

FROM SMALL BEGINNINGS

'If this suggestion is correct, the beginning of the world happened a little before the beginning of space and time.'
Georges Lemaître

It is almost impossible to comprehend the idea that the Universe we see today, one of infinite variety filled with billions of galaxies, trillions of stars and an inconceivable number of planets, harbouring an unimaginable variety of life, all came to be from a tiny space no bigger than the full stop at the end of this sentence. And yet it's an idea that we have seen grow from its first awkward conception in the early part of the twentieth century to the most comprehensive, evidence-based creation story in the history of human thought.

The invention of the term itself is credited to the English astronomer and science-fiction writer Fred Hoyle. Although Hoyle was one of the theory's most ardent critics, he was also the first to name it 'the Big Bang' during a BBC radio broadcast in 1948. Hoyle believed in a steady-state model of the Universe, a universe without a beginning or end, a theory that he held on to long after others had abandoned it. But despite this belief, it was his descriptive name that would go on to dominate our image of the Universe's beginning.

However, long before this title stuck, the theory itself had begun its slow rise to dominance. In the late 1920s, both Edwin Hubble and Belgian physicist and priest Georges Lemaître came up with the idea of an expanding universe. It was in fact Lemaître who published his finding first, describing what became known as Hubble's Law in an obscure paper that received minimal attention when it appeared in a little-known Belgian journal in 1927. As we've seen earlier in this chapter, it was Edwin Hubble who brought the concept of an expanding universe to the attention of the world two years later, casting Lemaître's work and name into the scientific shadows.

But Lemaître was far from finished. He reasoned that if the Universe was expanding simultaneously in every direction, that must mean yesterday the Universe was smaller than it is today, and the day before yesterday even smaller still. Follow this logic back day after day and across the eons of time, and the size of the Universe would shrink until eventually you get back to a single concentrated point from which the whole Universe would emerge. Publishing his thoughts in a short letter to *Nature* in May 1931, Lemaître captured in just over 450 words the first description of a 'world' that began with a 'primeval atom'. Published next to a letter describing the discovery of insect remains found in the gut of a cobra snake, Lemaître's brief note could have seemed inconsequential but it in fact sent a shockwave through the scientific community and was immediately picked up in the popular press around the world. The *New York Times* printed the whole of Lemaître's letter a few days later, and the now-famous 'Abbe Lemaître' was featured in an extensive article in *Popular Science* magazine. In a flurry of international interest, Lemaître travelled from the University of Louvain in Belgium to London under the invitation of his mentor, Arthur Eddington, who had taught Lemaître in Cambridge a few years earlier. The invitation included a chance to speak at the 100th anniversary meeting of the British Association for the Advancement of Science, which took place in the Great Hall of the University of London. The subject of the discussion was the relationship between the physical universe, spirituality and the evolution of the Universe, and Lemaître spared no time enthralling the distinguished audience with his new 'fireworks' theory of creation.

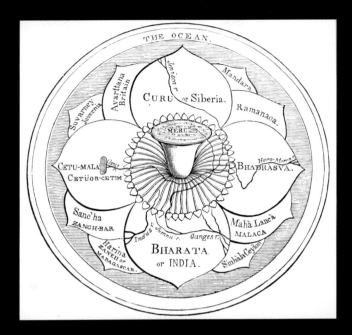

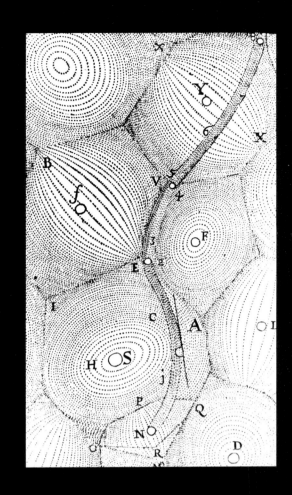

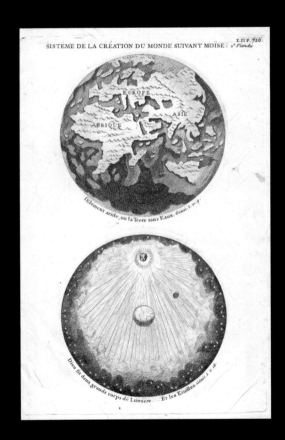

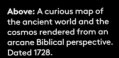

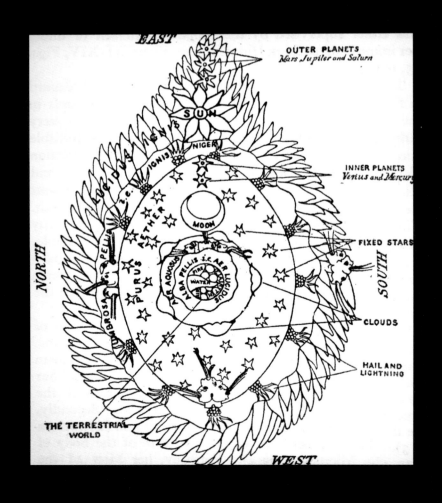

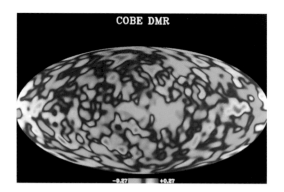

THE BEGINNING OF THE WORLD FROM THE POINT OF VIEW OF QUANTUM THEORY:

'Sir Arthur Eddington states that, philosophically, the notion of a beginning of the present order of Nature is repugnant to him. I would rather be inclined to think that the present state of quantum theory suggests a beginning of the world very different from the present order of Nature. Thermodynamical principles from the point of view of quantum theory may be stated as follows: (1) Energy of constant total amount is distributed in discrete quanta. (2) The number of distinct quanta is ever increasing. If we go back in the course of time we must find fewer and fewer quanta, until we find all the energy of the universe packed in a few or even in a unique quantum.

'Now, in atomic processes, the notions of space and time are no more than statistical notions; they fade out when applied to individual phenomena involving but a small number of quanta. If the world has begun with a single quantum, the notions of space and time would altogether fail to have any meaning at the beginning; they would only begin to have a sensible meaning when the original quantum had been divided into a sufficient number of quanta. If this suggestion is correct, the beginning of the world happened a little before the beginning

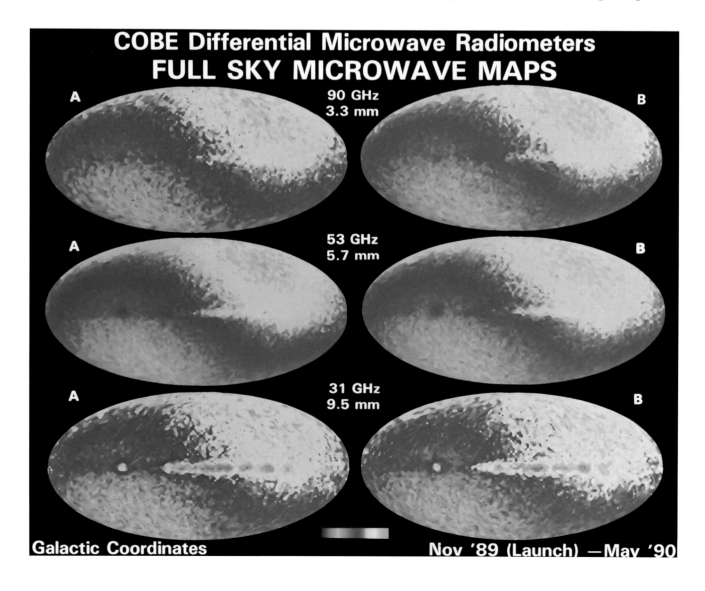

Right: Early microwave maps at 3 frequencies, colour-coded to show variations in the temperature of the microwave background.

Opposite: First microwave map of the whole sky showing ripples in the CMB. Produced using data from the Cosmic Background Explorer (COBE).

Below: DIRBE mapped the absolute sky brightness in 10 wavelength bands ranging from 1.25 microns to 240 microns to detect variation.

of space and time. I think that such a beginning of the world is far enough from the present order of Nature to be not at all repugnant. It may be difficult to follow up the idea in detail as we are not yet able to count the quantum packets in every case. For example, it may be that an atomic nucleus must be counted as a unique quantum, the atomic number acting as a kind of quantum number. If the future development of quantum theory happens to turn in that direction, we could conceive the beginning of the universe in the form of a unique atom, the atomic weight of which is the total mass of the universe. This highly unstable atom would divide in smaller and smaller atoms by a kind of super-radioactive process. Some remnant of this process might, according to Sir James Jeans's idea, foster the heat of the stars until our low atomic number atoms allowed life to be possible.

'Clearly the initial quantum could not conceal in itself the whole course of evolution; but, according to the principle of indeterminacy, that is not necessary. Our world is now understood to be a world where something really happens; the whole story of the world need not have been written down in the first quantum like a song on the disc of a phonograph. The whole matter of the world must have been present at the beginning, but the story it has to tell may be written step by step.'
Georges Lemaître, *Nature*

DIRBE Galactic Plane Maps

1.25 microns

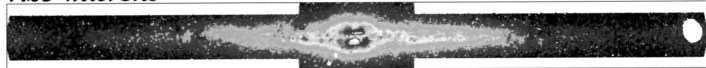

2.2 microns

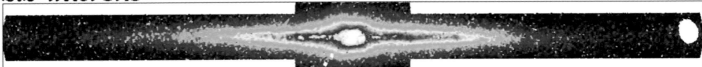

3.5 microns

4.9 microns

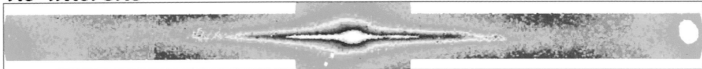

12 microns

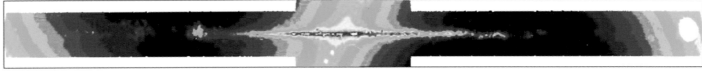

Below: Arno Penzias and Robert Wilson stand at the 15-metre Holmdel Horn Antenna that brought their most notable discovery.

Right: Meanwhile in Britain, the Jodrell Bank Telescope near Macclesfield, Cheshire, monitors radio emissions from individual objects in space.

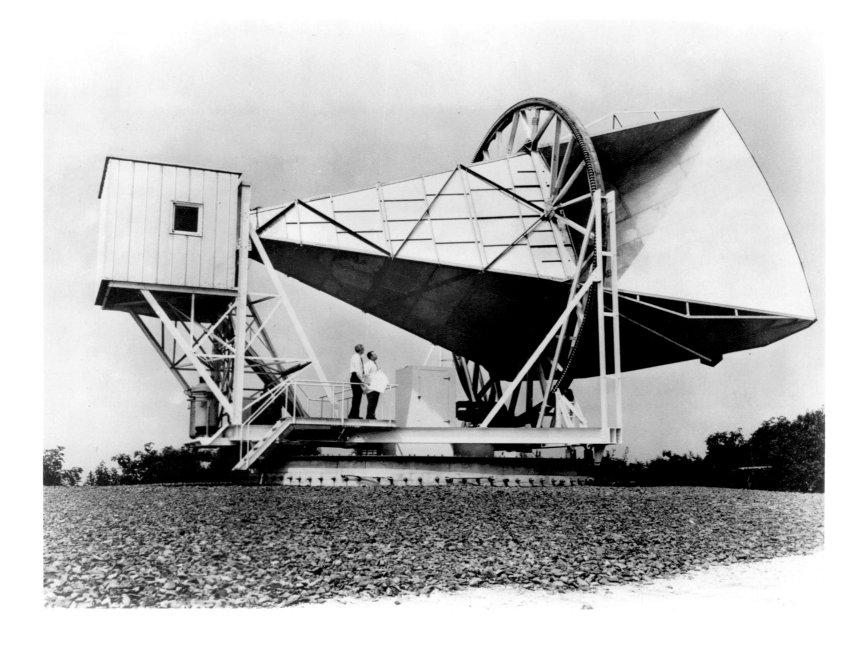

'Philosophically, the notion of
a beginning of the present order
of Nature is repugnant to me.'
Arthur Eddington

Above: Professor Martin Ryle
examines data recorded on
punched paper tape, converted
from radio receivers at Mullard
Radio Astronomy Observatory
in Cambridge.

Despite this brief moment of fame, Lemaître's theory remained outside of
the scientific mainstream and just one of a number of competing models put forth
to explain Hubble's observations. Even Einstein himself dismissed the Belgian's
work by supposedly telling Lemaître that his 'calculations were correct but his
physics was terrible'. With opponents of this standing, it was no surprise Lemaître's
primeval atom remained on the fringes. The world wasn't ready for a creation story
of such radical form, regardless of the evidence to support it, and it certainly wasn't
ready to accept it from a mathematician who also happened to be a priest. After
hundreds of years of scientific thought battling to move us away from a description
of a universe that begins with a moment of creation, it was perhaps too much to ask
for the world of science to accept a new empirical version of that story from a man
of God, no matter how good his maths. Commenting on this duality of approaches
many years later, Lemaître quipped, 'It appeared to me that there were two paths
to truth, and I decided to follow both of them.'

It wouldn't be until well after World War II that two clear frontrunners would
emerge in the race to explain the evidence of an expanding universe. On one side
was the steady-state model championed by Fred Hoyle, a universe without beginning
or end, where it was the creation of new matter that explained the expansion of the
Universe. On the other side was Lemaître's primeval atom hypothesis, by now termed
(by Hoyle) the Big Bang theory, which explained the expansion of the Universe in
terms of an ancient beginning; or, 'a kind of bottom in space and time,' as Lemaître
described it at a conference attended by all the leading figures in the field at Berkeley,
University of California, in August 1961. But while the debate raged through the
early 1960s, it was clear that the tide was only going in one direction, driven by
a growing body of evidence to support the Big Bang hypothesis. And then in one
moment, the game changed.

In 1964, Arno Penzias and Robert Wilson were working at the Bell Laboratories
in Holmdel, New Jersey, when they made an observation they couldn't explain.
Working with a highly sensitive, large microwave horn antenna, they'd picked up
a puzzling set of radio noise. Too weak to be a radio source coming from within the
Milky Way, it appeared to be coming not from a single source but simultaneously
from every direction in the sky, with the same intensity 24 hours a day and no change
across the passing of a year. With no explanation to match the observation, they
looked for a terrestrial source of the interference, exploring whether it could be radio
noise spilling out of New York City, even famously blaming bird faeces on the antenna
for the errant signal. But none of the potential explanations stood up, and with all
sources of interference rejected, it looked as if this strange radio signal would
remain unexplained. But in one of those remarkable coincidences, at the same time
that Penzias and Wilson were puzzling over their radiowaves, a team at Princeton
University, led by Robert Dicke, was predicting the theoretical existence of such
waves as a consequence of the Big Bang. While Dicke and his team rushed to build
their own antenna to see if they could detect the signal, word reached the team at
New Jersey and they soon realised they had the most profound explanation for the
strange radio signal. They had measured the afterglow of creation, the oldest light
in the Universe, born from the 'primordial flash' that had been stretched into
radiowaves by the vast expansion of the Universe. With this discovery we were
seeing further back in time than ever before; we'd finally broken through the darkness
and arrived at the beginning, at a 'day without a yesterday'. Or so we thought.

BEFORE THE BEGINNING

In the space of a single lifetime, we have journeyed further back into the story of our own creation than we could ever have dreamed possible. Back to a time before life began, before the Earth existed, before the Sun was born, before the Milky Way evolved into a home to billions of stars. Back ever further until we leave the first galaxy and the first star behind, back through the darkness until we arrive beyond the cosmic web and scaffold of the Universe, beyond the first atom to a time approximately 13.8 billion years ago when it all began. The beginning of time itself, when every part of everything in the Universe – including every part of you – was contained in an incredibly hot, incredibly dense point that has been expanding ever since.

By arriving here at what we have called the Big Bang, it seems as if we have reached the very farthest horizon, the beginning of space and time. But as Edwin Hubble suggested, 'the history of astronomy is the history of receding horizons', and even here on the edge of time itself, we now suspect there must be another horizon. We used to think that the Universe began in this very hot, very dense state at the very beginning of time, but we now strongly suspect that the Universe existed before that, and so in a very real sense it's now possible to speak of a time before the Big Bang.

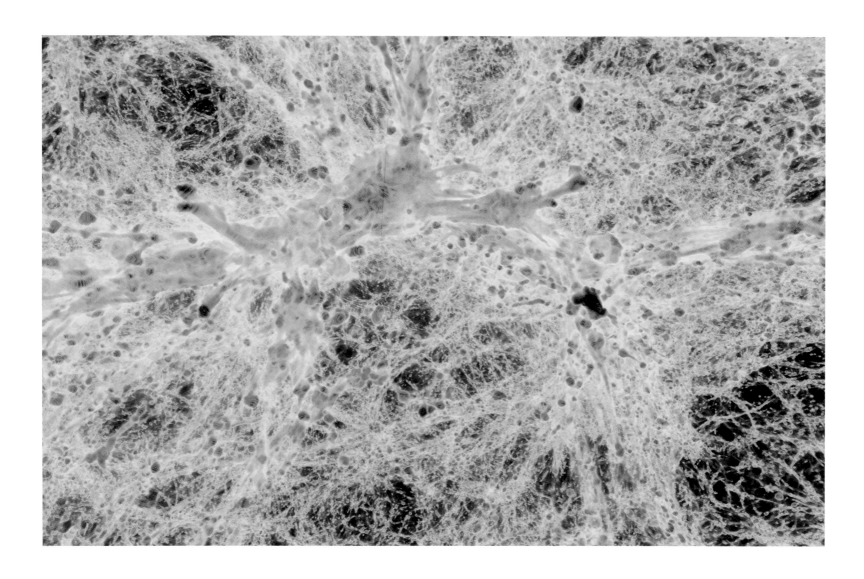

INFLATION

According to inflation theory, the Universe blew up so quickly that there was no time for the essential homogeneity to be broken, so the Universe after inflation would have been very uniform, even though parts of it were no longer in touch with each other.

Normal expansion

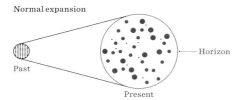

Past · Present · Horizon

Inflation

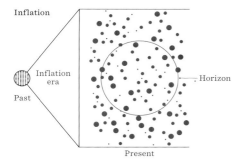

Past · Inflation era · Present · Horizon

Rapid expansion

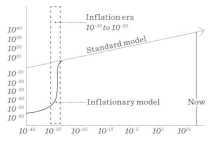

This journey back to a time before the Big Bang has been made possible by a powerful combination of theoretical and physical exploration. The theoretical journey began in the late 1970s when Alan Guth attended a lecture by Robert Dicke, the man who led the team that predicted the existence of an afterglow of radiowaves from the Big Bang. Dicke had provided some of the most fundamental evidence to underpin the Big Bang hypothesis, but there were still gaps in our understanding, characteristics of the Universe that could not be explained by the Big Bang hypothesis alone. Known as the 'flatness problem', Dicke and others were struggling to explain why the Universe had been born in such a uniform homogenous state, spatially flat, exactly the same temperature to an accuracy of 99.997 per cent in all directions and with no remnants of the ultra-high energies that existed at the beginning of time.

The solution Guth came up with was to take us back in time to a moment before the Big Bang. To a time and space inconceivably strange, where no matter existed, just colossal amounts of energy concentrated into the tiniest of specks. Nothing in this pre-Universe can connect to the cosmos we see today; it was a structureless void filled with nothing but energy, and yet something lurking within it was still able to trigger an unimaginably violent expansion that Guth named cosmic inflation.

According to Guth's theory, inflation made this primordial speck expand at a phenomenal rate, doubling in size 10 trillion trillion times every billionth of a second. The theory suggests that the period of inflation may have gone on for vast amounts of time – even for eternity – but this final period of exponential expansion continued for just a fraction of a second. We think it played out over just 10^{-32} seconds – that's a hundred million million million million millionth of a second, increasing the size of the Universe by a factor of around 10^{50}. That means it increased in size 100 million million million million million million million million times. No wonder Guth called this phenomenal expansion inflation.

As the inflationary period drew rapidly to a close, the Universe had expanded to somewhere between the size of a football and the size of a tower block. It was at this point that all the energy that had filled the early Universe, enough to make 2 trillion galaxies each full of billions upon billions of stars and planets, went from driving the rapid expansion to being transformed into the most basic ingredients that make up everything in our observable universe. And it is that moment – the moment of transformation from a universe of pure expanding energy to a universe expanding far more slowly with all the primordial ingredients in place to build the Universe that we see today – that we call the Big Bang.

So the Big Bang was not, as Lemaître first imagined, some kind of enormous 'burst of fireworks', some kind of explosion. It was actually a transformation, a transformation of energy into matter, a moment when that speck – the pre-Universe – became our universe.

Opposite: Supercomputer model of radiation hydrodynamics in galaxies in the early Universe, with over 1 billion dark matter and star particles.

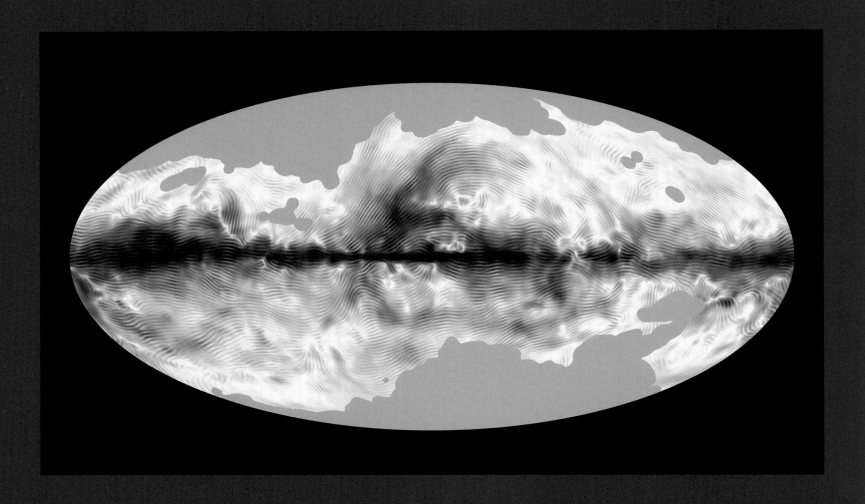
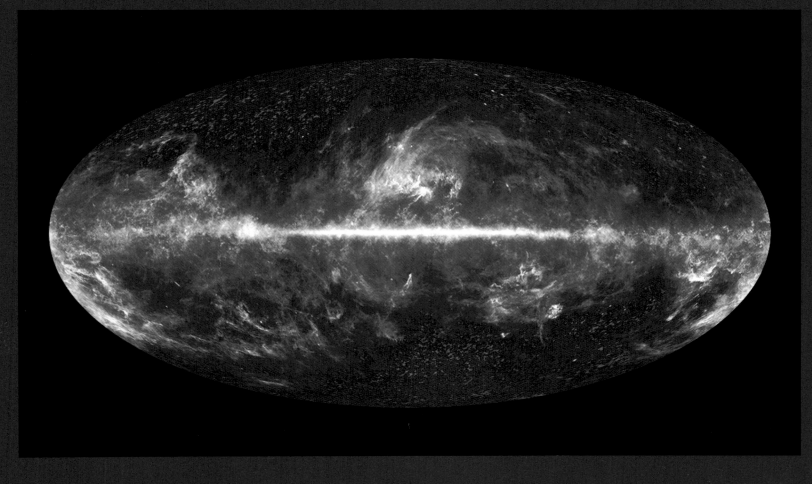

CLUES IN THE DARK

The idea of inflation is so beyond the realm of our experience of the Universe in which we live today, that it is almost impossible to imagine how we can transport ourselves back to that moment and find evidence to support the theory. But amazingly, over the last 30 years we have been able to peer back and discover direct evidence to support the theoretical model of inflation.

To understand how we have gone about searching for this evidence, we have to go back to the discovery that underpinned the theory of the Big Bang in the first place. In 1964, when Penzias and Wilson detected the strange radio signal that was shown to be the direct afterglow of the Big Bang, it triggered decades of research into the detail that could be detected within that first light. And this exploration has enabled us to search for clues within the light that led right back to those first pivotal moments in the history of the Universe, the very moments between the inflationary period coming to an end and the Big Bang beginning.

Today, we call the primeval fireball that Penzias and Wilson first detected the Cosmic Microwave Background (CMB), and the images we have been able to take of this 'oldest light' have helped transform our understanding of the early Universe. The light contained within the CMB has been stretched as it's travelled across the Universe for 13.8 billion years; it's been redshifted so far it is now reaching us as microwaves, some of the longest wavelengths of light in the electromagnetic spectrum. The light itself emanated from a moment in the Universe's history approximately 380,000 years after the Big Bang. It was at this moment that our calculations suggest the Universe became cool enough for the electrons and protons to form into hydrogen atoms, creating the conditions for the Universe to become transparent to light for the first time. So light was able to travel without being scattered by the early fog of free electrons and other subatomic particles that filled the Universe before this time. The light itself is not from any stars or galaxies, as this is a time long before any of those came into existence; instead the light of the CMB is given off by the heat of the early Universe, glowing like the 'embers' Lemaître imagined would be left after that first moment of creation. So it really is the afterglow of the Big Bang, visible only after the Universe has cooled down enough to let it escape.

Over successive decades, we have been able to image this 'baby picture' of the Universe in ever more detail using a series of increasingly powerful telescopes. The first of these was the Cosmic Background Explorer (COBE) satellite that was launched by NASA in 1989, which gave us our first real glimpse of the CMB in 1992 down to a resolution of about seven degrees. This was followed by the Wilkinson Microwave Anisotropy Probe (WMAP), launched in 2001, which pushed the imaging of the CMB even further, to a resolution of about half a degree.

But this was all about to be blown away by the extraordinary power of the Planck Telescope. Launched by the European Space Agency on 14 May 2009, its mission trajectory placed it into a precise position 1.5 million kilometres from Earth, ready to begin its measurement of the CMB with exquisite accuracy. Carrying two highly sensitive instruments built to detect the ancient light, Planck was designed with a cooling system that allowed the telescope to have a functioning temperature of -273.05 degrees Celsius (just 0.1 degrees above absolute zero), making it the coldest known object in the Universe until the cryocooler ran out of power. This innovative design allowed Planck to operate with such sensitivity that the limits of what it can see are not constrained by its own instruments, rather by the fundamental physics

COMPOSITION OF THE COSMOS
Before and after Planck's 2013 data release. These adjustments largely came from an increase in the trustworthiness of power spectrum at smaller scales, where the effect of dark matter became more important.

Before Planck

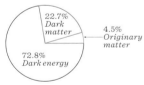

22.7%
Dark matter

4.5%
Originary matter

72.8%
Dark energy

After Planck

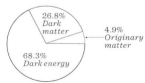

26.8%
Dark matter

4.9%
Originary matter

68.3%
Dark energy

Opposite: Microwave sky with tiny temperature fluctuations (bright spots) reflecting density variations from which the cosmic web of our universe originated.

Opposite top: Polarised light emitted by interstellar dust in the Milky Way shows the Milky Way's magnetic field.

Technology has really brought the astronomy
community together to build some of the most
astounding machines out there, and we are really
pushing the limits in terms of what we can discover –
the biggest telescopes, the most fantastically complex
systems, and we get to use them to understand the
Universe. Astronomy has always been at the forefront
of technology, but the kind of technology we work
with right now is absolutely amazing. And it's really
told us we have a completely new picture of the
Universe than what we have had for almost the
entirety of human history. And that's really special.
It's quite humbling.
David DeSario, Astronomer, Durham University

Opposite: The protective
cover is removed from the
Planck spacecraft in a clean
room at Europe's launch site
in Kourou, French Guiana.

Above: Planck's primary
mission is to study the Cosmic
Microwave Background,
the radiation left over from
the Big Bang.

of the CMB light itself. This means that the image of the CMB that Planck created
at a resolution of 0.07 degrees will very possibly never be bettered, but it also means
that the image first released in 2013 provides us with the ultimate window into
the early history of the Universe and the inflationary period that started it all off.

The photograph of the CMB that Planck provided has enabled us to peer
further into the detail of the distant past than ever before. Capturing light that
has travelled for almost 13.8 billion years before reaching us, it's a photograph of
the entire sky, the celestial sphere laid flat so that we can see it in its entirety – an
almost featureless glow from a time before starlight when the Universe itself was
all that shone. But contained within this image is an extraordinary level of detail,
because hidden in the colour is a pattern of ripples that holds within it clues to the
origin of the Universe.

To understand these clues hidden in the CMB, we first have to return to the
moment of its creation 380,000 years after the Big Bang. At this precise moment,
the photons of light that Planck has captured in the image of the CMB were
suddenly released from the primordial fog of the early Universe. Up to this point,
photons and other subatomic particles had been so tightly coupled together that
they formed a single 'fluid' of matter and radiation, but as the temperature of the
Universe dropped, these two components split apart from each other in a process
known as decoupling, setting the first photons free to begin their journey across
the Universe, and ultimately, after a 13.8-billion-year journey, into the detectors of
the Planck Telescope. Crucially, this ancient light was not just released evenly at
that critical moment; each photon carries a memory of the structure of the Universe
it was once trapped inside, revealing to us the distribution of matter and radiation
at that moment of transformation.

Such detail is revealed not directly in the CMB light but indirectly through the
temperature fluctuations we see across it. The reason these fluctuations are so
revealing is because the temperature at any point on the CMB image relates directly
to the structure of that part of the Universe at that time. So, if a photon was in a
slightly denser part of space at the time they were set free, they would have had
to expend a little more energy to break away and start their long journey across the
Universe, meaning that they would appear to be slightly cooler than photons that
were located in a slightly less dense region of space. In this way, the temperature
fluctuations in the CMB that Planck has been able to chart so accurately give
us a map that also reveals the minute variations in the density of matter across
the Universe at that time, the structure of our universe at a time before galaxies,
stars or planets existed. The CMB is the ultimate snapshot of the structure of
our universe as it was just a few hundreds of thousands of years after it all began.

It's for this reason that this pattern in the CMB is seen by many as one of the most
important discoveries in all of human history, because it reveals the evolution of the
Universe at a critical moment in its history. Wind time forwards from this moment
and we see that those fluctuations in the density of the amorphous universe are
directly linked to the Universe that we see today. That's because those fluctuations
in the density of matter in the early Universe persisted, then the areas of slightly
denser matter gradually grew, increasing in density as their gravitational influence
strengthened. They then drew in more and more matter until those primordial
fluctuations in density became the seeds from which the Universe developed,

Right: Whole-sky image from Planck of the Cosmic Microwave Background, the relic radiation from the Big Bang billions of years ago.

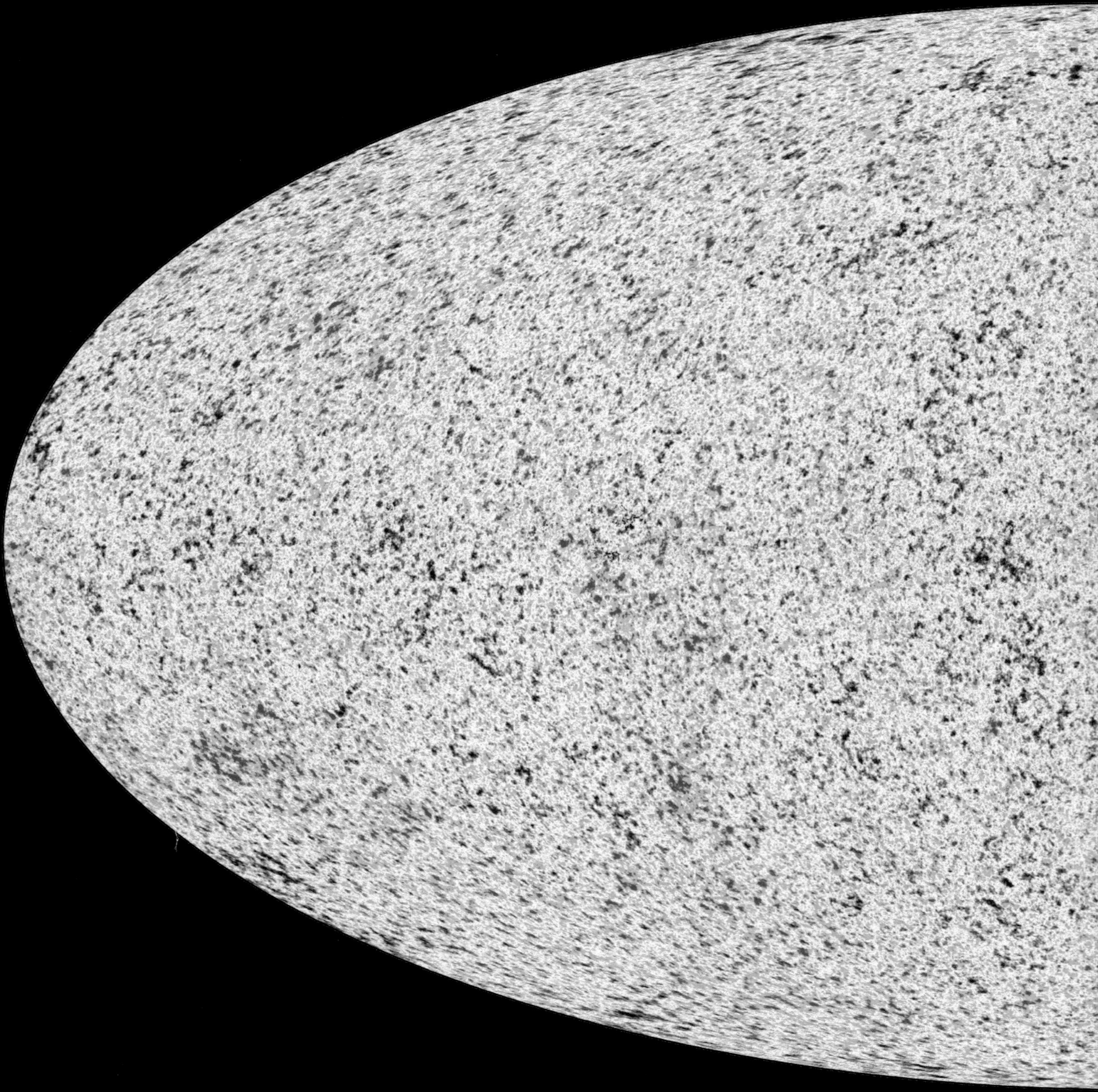

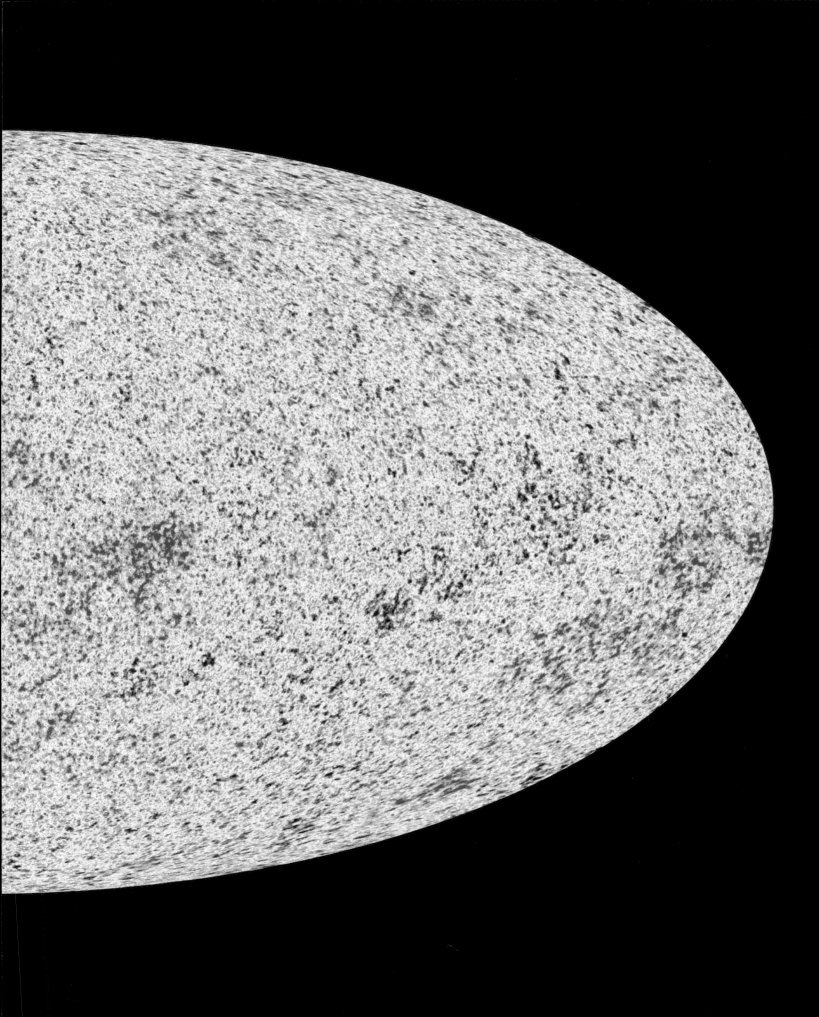

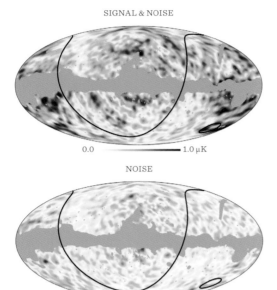

SIGNAL & NOISE

0.0   1.0 µK

NOISE

0.0   1.0 µK

condensing first in the form of the cosmic web, followed by all the stars and galaxies that we see today. All of it – every one of the 2 trillion galaxies that we think may populate the Universe today and all of the stars and planets within them, including our own – can be traced back to the structure we see in the CMB.

But that's not the end of the power of this image, because not only does it give us a starting point in the Universe's history to roll forwards from, it also gives a moment from which we can turn back the clock. Those fluctuations in density that we see in the CMB and that gave rise to the Universe we see today also had to come from somewhere, so as the clock ticks backwards in time we can follow those structures as they get smaller and smaller until we find ourselves at a time before the Big Bang, into that moment when the Universe was about to undergo the most rapid inflation.

It's here, in a universe no bigger than a speck, filled with all the energy needed to create a whole universe filled with stars, that we find the first trace of that ancient map, the structures from which everything else would follow. The idea that these primordial fluctuations existed in that early ocean of energy that drove inflation was first suggested by a group of theoretical physicists working on the inflation model back in the early 1980s. Scientists including Alan Guth and Stephen Hawking theorised that inflation could not have taken place within a perfectly still ocean of energy (and so would not produce a perfectly symmetrical universe). Instead, the Universe prior to the rapid inflationary period would have contained tiny quantum fluctuations within it, ripples in the ocean of energy from which the Universe would

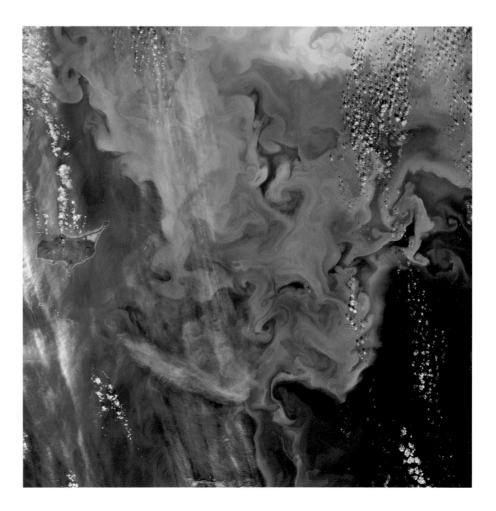

Top: Polarisation maps comparing Planck data from both signal and noise versus noise indicate some as yet unexplained anomalies.

Right: The early Universe can be compared to an ocean of energy, and the quantum ripples are the variations that will become all matter.

Below: The early Universe was filled with a liquid plasma of subatomic particles, so it's useful to imagine forces like ocean currents at work.

be built. Theoretical predictions are one thing, but the fluctuations in the density of the CMB give us direct evidence of the existence of those quantum fluctuations. This is because these fluctuations were imprinted on the Universe as inflation came to an end and the Universe transformed from pure energy into a space full of matter, matter that was imprinted with a structure that created areas of slightly different density. Thus were sown the primordial seeds of everything that was to come.

So this is our creation story. A story that begins with a speck of unimaginable energy, a tiny seed from which a whole universe would grow. But this was no perfect seed – imperfections in its structure, ripples in this ocean of energy, would appear as a consequence of nature. And so, as the inflationary epoch drove the briefest but most violent expansion, the ripples in the speck stretched with it, imprinting themselves into the structure of the Universe as inflation ended and energy turned into matter. We have seen those tiny fluctuations of density in the Universe's first matter growing and thickening into regions of space in the shadows cast by the most ancient of light. Out of these clumps of matter the first stars grew, then the first galaxies, all following the blueprint laid down long before. A blueprint that created one galaxy among 2 trillion islands of stars within which, after 9 billion years, a star was formed and eight planets emerged. And on one of those planets, those tiny fluctuations laid down 13.8 million years prior gave rise to a collection of atoms that could look and think and explore how and why the Universe and everything within it had come to be.

'Planck is capturing the earliest light that's ever been available in our universe; tiny little unevenness in the earliest moments in cosmic history that would grow and grow to form the cosmic web around us.'
David Kaiser, Physicist, MIT

To end, let's revisit the words of Professor Brian Cox:

'What does it mean to be human? Why do we exist? Why does anything exist at all? These do not sound like scientific questions, they sound like questions for philosophy or theology. But I think they are scientific questions because they are questions about nature. They are questions about the Universe.

'The way we understand the Universe is by observing it. We've seen ripples in the most ancient light in the Universe that were caused by events that happened before the Big Bang. We've seen streams of galaxies, billions of them written across the sky in a giant cosmic web. We've seen thousands of planets orbiting around distant stars, worlds beyond imagination. The lesson for me is clear: we won't answer the deepest questions by being introverted, by looking inward, we will answer them by lifting our gaze above the horizon and looking outwards into the Universe beyond the stars. We used to look to the sky and see only questions. Now we are beginning to see answers.'

INDEX

PICTURE CREDITS
t: top, b: bottom, l: left, r: right, m: middle

All reasonable efforts have been made to trace the copyright owners of the images in this book. In the event that there are any mistakes or omissions, updates will be incorporated for future editions.

1 Forgem / Shutterstock; 2 NASA, ESA and J Maíz Apellániz (Instituto de Astrofísica de Andalucía); 5 Forgem / Shutterstock; 7 NASA / Joel Kowsky; 8 SCIEPRO / SCIENCE PHOTO LIBRARY; 10 NASA, ESA and STScI; 13 NASA / JPL-CALTECH / SSI / CORNELL / SCIENCE PHOTO LIBRARY; 14 tl CPA Media Pte Ltd / Alamy Stock Photo, bl ARABIC MANUSCRIPTS COLLECTION / NEW YORK PUBLIC LIBRARY / SCIENCE PHOTO LIBRARY, r Bartolomeu Velho, 1568, Public Domain, Wikimedia Commons; 15 t The Granger Collection / Alamy Stock Photo, b Giordano Bruno, 1588, Public Domain, Wikimedia Commons; 16 tl NASA Photo / Alamy Stock Photo, ml Science History Images / Alamy Stock Photo, tr Science History Images / Alamy Stock Photo, mr NASA / VRS / DETLEV VAN RAVENSWAAY / SCIENCE PHOTO LIBRARY, br 504 collection / Alamy Stock Photo; 17 Chronicle / Alamy Stock Photo; 18 DAVID PARKER / SCIENCE PHOTO LIBRARY; 19 EUROPEAN SOUTHERN OBSERVATORY / SCIENCE PHOTO LIBRARY; 20 t NASA / SCIENCE PHOTO LIBRARY, b NASA / JPL / SCIENCE PHOTO LIBRARY; 21 l NASA / GODDARD SPACE FLIGHT CENTER / SDO / SCIENCE PHOTO LIBRARY, r NASA / GODDARD SPACE FLIGHT CENTER / SDO / SCIENCE PHOTO LIBRARY; 22 Dotted Yeti / Shutterstock; 23 t Leire P J / DYDPPA / Shutterstock, b Hemis / Alamy Stock Photo; 25 NASA / ESA; 26 NASA / JPL-Caltech; 27 Frances Roberts / Alamy Stock Photo; 28 tl NASA / Kim Shiflett, bl NASA / Chris Rhodes, tr NASA / Jim Grossmann, m NASA / Kim Shiflett, br NASA / Jim Grossmann; 29 l LYNETTE COOK / SCIENCE PHOTO LIBRARY, r Andrew Z Colvin, Public Domain, Wikimedia Commons; 31 t NASA / JPL-Caltech, b NASA / Ames / JPL-Caltech; 32 Harvard-Smithsonian Center for Astrophysics / David Aguilar; 33 JLStock / Shutterstock; 34 ADRIAN BICKER / SCIENCE PHOTO LIBRARY; 35 Gareth McCormack / Alamy Stock Photo; 36 Huw Griffiths / British Antarctic Survey; 37 t Anup Shah / naturepl.com, bl Image by Joshua Stevens, using Landsat data from the US Geological Survey, br GUDKOV ANDREY / Shutterstock; 38 NASA / SCIENCE PHOTO LIBRARY; 39 KEES VEENENBOS / SCIENCE PHOTO LIBRARY; 40 PETER FRETWELL / PETE BUCKTROUT, BRITISH ANTARCTIC SURVEY / NASA / GSFC / METI / JAPAN SPACE SYSTEMS / SCIENCE PHOTO LIBRARY; 41 Morley Read / naturepl.com; 42 NASA / JPL-Caltech; 42 inset NASA / JPL-Caltech; 43 l NASA /JPL-Caltech / ASU / MSSS, r NASA / JPL-Caltech, b NASA / JPL-Caltech; 44 ENRIQUE LOPEZ-TAPIA / NATURE PICTURE LIBRARY / SCIENCE PHOTO LIBRARY; 45 NASA/JPL-Caltech; 46 t NASA / Ames Research Center / Wendy Stenzel, bl ESA / ROSETTA / MPS FOR OSIRIS TEAM MPS / UPD / LAM / IAA / SSO / INTA / UPM / DASP / IDA; NAVCAM: ESA / ROSETTA / NAVCAM / SCIENCE PHOTO LIBRARY; 47 NASA / Ames / SETI Institute / JPL-Caltech; 48 Paulo Afonso / Shutterstock; 49 RGB Ventures / SuperStock / Alamy Stock Photo; 50 Jiann / Shutterstock; 51 t NASA image by Jeff Schmaltz, MODIS Rapid Response Team, Goddard Space Flight Center, b BorneoRimbawan / Shutterstock; 52 SCIENCE PHOTO LIBRARY; 53 tl NASA / Chris Gunn, tm NASA / Northrop Grumman, tr NASA / Goddard Space Flight Center / Laura Betz, b NASA / Alex Evers; 54 NASA / JPL-Caltech; 55 J Marshall Tribaleye Images / Alamy Stock Photo; 56 t NASA, ESA, and G Bacon (STScI), b ESA,D DUCROS / SCIENCE PHOTO LIBRARY; 57 l GREGOIRE CIRADE / SCIENCE PHOTO LIBRARY, r EUROPEAN SOUTHERN OBSERVATORY / SCIENCE PHOTO LIBRARY; 58 NASA /JPL-Caltech; 61 NASA / JPL-Caltech / GSFC / JAXA; 62 Jack Dykinga / naturepl.com; 64 Science History Images / Alamy Stock Photo; 65 The Print Collector / Alamy Stock Photo; bl 66 Chronicle / Alamy Stock Photo, br Prof. Peter Fowler / Science Photo Library; 67 bl EFDA-JET / SCIENCE PHOTO LIBRARY , br Royal Astronomical Society / Science Photo Library; 68 World History Archive / Alamy Stock Photo; 69 Smithsonian Institution Archives; 70 NASA / JPL / California Institute of Technology; 71 ROYAL ASTRONOMICAL SOCIETY / SCIENCE PHOTO LIBRARY; 72 Rashevskyi Viacheslav / Shutterstock, inset GREGOIRE CIRADE / SCIENCE PHOTO LIBRARY; 73 t NASA / GSFC / SDO, m Geopix / Alamy Stock Photo, b DAMIAN PEACH / SCIENCE PHOTO LIBRARY; 74 ESA and the Planck Collaboration / H Dole, D Guéry & G Hurier, IAS / Univ. Paris-Sud / CNRS / CNES; 75 NASA, ESA and M Montes (Univ. of New South Wales); 76 ESA / Hubble & NASA; 79 l Hideki Umehata, r Joshua Borrow using C-EAGLE; 80 t Volker Springel / Max Planck Institute for Astrophysics / Science Photo Library, b Pavel_Klimenko / Shutterstock; 81 M Najjar / ESO; 83 tl Everett Collection Historical / Alamy Stock Photo, tr Everett Collection / Shutterstock, b Everett Collection / Shutterstock; 84 Suzyj, Public Domain, Wikimedia Commons; 85 NASA / ESA / STSCI / R SCHALLER / SCIENCE PHOTO LIBRARY; 86 row 1 l MICHAEL W DAVIDSON / SCIENCE PHOTO LIBRARY, row 1 m EDWARD KINSMAN / SCIENCE PHOTO LIBRARY, row 1 r KTSDESIGN / SCIENCE PHOTO LIBRARY, row 2 l GERD GUENTHER / SCIENCE PHOTO LIBRARY, row 2 m POWER AND SYRED / SCIENCE PHOTO LIBRARY, row 2 r DIRK WIERSMA / SCIENCE PHOTO LIBRARY, row 3 l ANDRE GEIM, KOSTYA NOVOSELOV / SCIENCE PHOTO LIBRARY, row 3 m DENNIS KUNKEL MICROSCOPY / SCIENCE PHOTO LIBRARY, row 3 r MASSIMO BREGA / SCIENCE PHOTO LIBRARY, row 4 l Rich Carey / Shutterstock, row 4 m Andrzej Kubik / Shutterstock, row 4 r Albert Beukhof / Shutterstock; 88 NASA / JPL-Caltech / ESA / CXC / Univ. of AZ/Univ of Szeged; 89 DSS, STScI / AURA, Palomar / Caltech and UKSTU / AAO; 90 NASA, ESA, N Smith (Univ. of California, Berkeley), and The Hubble Heritage Team (STScI / AURA); 92 NASA / GSFC / M Corcoran et al; 93 l NASA / JPL-Caltech / Harvard-Smithsonian CfA, tr NASA / JPL-Caltech, br NASA, ESA, and The Hubble Heritage Team (STScI/AURA); 94 NASA, ESA, AND THE HUBBLE HERITAGE TEAM (STSCI / AURA) / SCIENCE PHOTO LIBRARY; 95 NASA, ESA, STScI, N Habel and S T Megeath (Univ. of Toledo); 96 NASA Goddard; 97 TWStock / Shutterstock; 98 tl AC NewsPhoto / Alamy Stock Photo, tr Earth Science and Remote Sensing Unit, NASA Johnson Space Center, bl Science History Images / Alamy Stock Photo, br UPI / Alamy Stock Photo; 99 tl NASA / JPL-Caltech / Univ. of Arizona, tr Nasa / JPL-Caltech / Univ. of Arizona / Science Photo Library, bl NASA / JPL-Caltech / MSSS, br NASA / JPL-Caltech / Univ. of Arizona; 100 NASA / Ames Research Centre; 101 tl Granger Historical Picture Archive / Alamy Stock Photo, m NASA/JPL, r NG Images / Alamy Stock Photo; 102 Volgi archive / Alamy Stock Photo; 103 NASA/SDO; 104 NASA / Naval Research Laboratory / Parker Solar Probe / Science Photo Library; 105 tl NASA/Kim Shiflett, tm NASA / Leif Heimbold, r NASA / Kim Shiflett, b NASA / Bill Ingalls; 106 Yvonne Baur / Shutterstock; 108 t NOAA / C German (WHOI), b John Durham / Science Photo Library; 110 NASA / JPL-Caltech; 111 NASA / JPL-Caltech; 112 t NASA / CXC / M Weiss, b NASA/JPL-Caltech); 113 NASA / JPL-Caltech / ESA; 115 NASA / ESA / STSCI / HUBBLE HERITAGE TEAM / SCIENCE PHOTO LIBRARY; 117 t NASA / JPL-Caltech, m NASA, ESA, the Hubble Heritage Team (STScI/AURA)-ESA/ Hubble Collaboration and M Stiavelli (STScI), b ESA / Hubble and NASA; Acknowledgement: Judy Schmidt; 118 t Sunyaev-Zel'dovich effect: ESA Planck Collaboration; optical image: STScI Digitized Sky Survey, b NASA/JPL-Caltech / AMES /Univ. of Birmingham; 119 t MIKKEL JUUL JENSEN / SCIENCE PHOTO LIBRARY, b NASA / JPL-Caltech / Univ. of Arizona; 120 Ron Miller / Stocktrek Images; 121 l NASA / ESA / JPL / Q D WANG / S STOLOVY / SCIENCE PHOTO LIBRARY, r ESA/Herschel / PACS / SPIRE / Ke Wang et al. 2015; 122 Universal History Archive / UIG / Bridgeman Images; 123 Petr Kratochvila / Shutterstock; 124 thipjang / Shutterstock; 126 CORVAJA / ESA / SCIENCE PHOTO LIBRARY; 127 l ESA / ATG medialab, tr M PEDOUSSAUT / ESA / SCIENCE PHOTO LIBRARY, br M PEDOUSSAUT / ESA / SCIENCE PHOTO LIBRARY; 128 Lost in Time / Shutterstock; 130 ROBERT GENDLER / SCIENCE PHOTO LIBRARY; 131 t ThomasLENNE / Shutterstock, b picturepixx / Shutterstock; 132 t Pichet siritantiwat / Shutterstock, b NASA/ESA / JPL-Caltech / Conroy et. al. 2021; 133 t ESA / Planck Collaboration, b ESA / NASA / JPL-Caltech; 134 ESA / GAIA / DPAC / SCIENCE PHOTO LIBRARY; 136 NASA, ESA, the Hubble Heritage (STScI/ AURA)-ESA / Hubble Collaboration and A Evans (Univ. of Virginia, Charlottesville / NRAO / Stony Brook University); 138 JUAN CARLOS CASADO (STARRYEARTH.COM) / SCIENCE PHOTO LIBRARY; 139 tl The History Collection / Alamy Stock Photo, tr Science History Images / Alamy Stock Photo, b Art Collection 2 / Alamy Stock Photo; 140 ESA / Hubble & NASA; 141 ESA; 142 ADAM BLOCK / MOUNT LEMMON SKYCENTER / UNIV. OF ARIZONA / SCIENCE PHOTO LIBRARY; 144 t LWEINSTEIN, NASA / SCIENCE PHOTO LIBRARY, b NASA / ESA / STSCI / SCIENCE PHOTO LIBRARY; 146 DR RUDOLPH SCHILD / SCIENCE PHOTO LIBRARY; 147 tl ROYAL ASTRONOMICAL SOCIETY / SCIENCE PHOTO LIBRARY, bl ROYAL ASTRONOMICAL SOCIETY / SCIENCE PHOTO LIBRARY, r ROYAL ASTRONOMICAL SOCIETY / SCIENCE PHOTO LIBRARY; 148 t NASA / JPL-Caltech / UCLA, b NASA / JPL-Caltech; 150 ROYAL ASTRONOMICAL SOCIETY / SCIENCE PHOTO LIBRARY; 151 t FLHC 50 / Alamy Stock Photo, m Science History Images / Alamy Stock Photo, b National Academy of Sciences, nasonline.org; 152 l Public Domain, Wikimedia Commons, r SOTK2011 / Alamy Stock Photo; 153 t NASA Goddard, b ECKHARD SLAWIK / SCIENCE PHOTO LIBRARY; 154 NASA, ESA, J Dalcanton (Univ. of Washington), B F Williams (Univ. of Washington), L C Johnson (Univ. of Washington), the PHAT team, and R Gendler; 156 tl NASA, ESA, J DePasquale and E Wheatley (STScI) and Z Levay, tr ESA / Gaia (star motions); NASA / Galex (background image); R van der Marel, M Fardal, J Sahlmann (STScI), b NASA; 157 Nasa; 159 NASA; ESA; Z Levay and R van der Marel, STScI; T Hallas and A Mellinger; 161 NASA / JPL-Caltech; 162 DAVID A HARDY, FUTURES: 50 YEARS IN SPACE / SCIENCE PHOTO LIBRARY; 163 PROF YOSHIAKI SOFUE / SCIENCE PHOTO LIBRARY; 164 NASA / CXC / Amherst College / DHaggard et al; 166 NASA / Marshall Space Flight Center; 167 tl Marshall Space Flight Center collection, tr NASA Goddard, bl NASA / CXC / JPL-Caltech / PSU / CfA, br NASA Goddard; 168 bennu phoenix / Alamy Stock Photo; 169 Everett Collection Historical / Alamy Stock Photo; 170 t SCIENCE PHOTO LIBRARY; 170 b NASA / JPL-Caltech / SSC; 171 NASA Goddard; 173 NASA Goddard; 174 t Public Domain, Wikimedia Commons, b NASA, ESA, AND D PLAYER (STSCI) / SCIENCE PHOTO LIBRARY; 176 NASA / JPL-Caltech/Novapix / Bridgeman Images; 177 NASA'S GODDARD SPACE FLIGHT CENTER / JEREMY SCHNITTMAN / SCIENCE PHOTO LIBRARY; 178 NASA / Marshall Space Flight Center; 179 Gado Images / Alamy Stock Photo; 180 MARK GARLICK / SCIENCE PHOTO LIBRARY; 182 NASA / Marshall Space Flight Center; 183 t NASA Image Collection / Alamy Stock Photo, b Xinhua/Shutterstock; 184 NASA, ESA, and D Coe, J Anderson and R van der Marel (Space Telescope Science Institute); Acknowledgement for Omega Centauri Image: NASA, ESA and the Hubble SM4 ERO Team; Science: NASA, ESA, C-P Ma (Univ. of California, Berkeley), and J Thomas (Max Planck Institute for Extraterrestrial Physics, Garching, Germany); 186 EHT COLLABORATION / EUROPEAN SOUTHERN OBSERVATORY / SCIENCE PHOTO LIBRARY; 187 Matteo Omied / Alamy Stock Photo; 188 X-ray: NASA / CXC / Villanova University / J Neilsen; Radio: Event Horizon Telescope Collaboration; 189 EUROPEAN SOUTHERN OBSERVATORY / EHT COLLABORATION / SCIENCE PHOTO LIBRARY; 190 The Print Collector / Alamy Stock Photo; 191 t Hezekiah Conant, US Patent 371306 A, Public Domain, Wikimedia Commons, b John D Sirlin / Shutterstock.com; 192 t ALMA (ESO / NAOJ / NRAO) / JR Goicoechea (Instituto de Física Fundamental, CSIC); 192 b Richard Wainscoat / Alamy Stock Photo; 193 L CALCADA / SPACEENGINE.ORG / EUROPEAN SOUTHERN OBSERVATORY / SCIENCE PHOTO LIBRARY; 194 NASA, ESA, S Baum and C O'Dea (RIT), R Perley and W Cotton (NRAO / AUI / NSF), and the Hubble Heritage Team (STScI / AURA); 196 b NASA / SCIENCE PHOTO LIBRARY, tl NASA / SCIENCE PHOTO LIBRARY, tr NASA / Kim Shiflett; 197 NASA Goddard; 198 NASA / DOE / FERMI LAT COLLABORATION / SCIENCE PHOTO LIBRARY; 199 t NASA / JPL-Caltech, b NASA / JPL-Caltech; 200 NASA AND G BACON (STSCI) / SCIENCE PHOTO LIBRARY; 202 MARK GARLICK / SCIENCE PHOTO LIBRARY; 203 M PARSA / L CALCADA / EUROPEAN SOUTHERN OBSERVATORY / SCIENCE PHOTO LIBRARY; 204 RICHARD KAIL / SCIENCE PHOTO LIBRARY; 206 NASA / SCIENCE PHOTO LIBRARY; 209 NASA / JPL-Caltech; 210 ALFREDO RUIZ HUERGA / Alamy Stock Photo; 212 MIKKEL JUUL JENSEN / SCIENCE PHOTO LIBRARY; 213 PLANCK COLLABORATION / ESA / SCIENCE PHOTO LIBRARY; 214 NASA Image Collection / Alamy Stock Photo; 215 t Heritage Image Partnership Ltd / Alamy Stock Photo, b J Marshall Tribaleye Images / Alamy Stock Photo; 216 NASA / JPL-CALTECH / SCIENCE PHOTO LIBRARY; 217 l Nerthuz / Alamy Stock Photo, r NASA / ESA / HUBBLE / LEO SHATZ / SCIENCE PHOTO LIBRARY; 218 NASA, ESA, H TEPLITZ AND M RAFELSKI (IPAC / CALTECH), A KOEKEMOER (STSCI), R WINDHORST (ARIZONA STATE UNIVERSITY), AND Z LEVAY (STSCI) / SCIENCE PHOTO LIBRARY; 219 tl NC Collections / Alamy Stock Photo, tr LWM / NASA / Alamy Stock Photo, b NASA / Alamy Stock Photo; 220 NASA, ESA and P Oesch (Yale); 221 Brad Mitchell / Alamy Stock Photo; 222 RGB Ventures / SuperStock / Alamy Stock Photo; 223 NASA, ESA, R Ellis (Caltech), and the HUDF 2012 Team; 224 b Anonymous / AP / Shutterstock; 225 Granger / Shutterstock; 226 t Matteo Omied / Alamy Stock Photo, b ESA / GAIA / DPAC / SCIENCE PHOTO LIBRARY; 227 VVV CONSORTIUM / D MINNITI / EUROPEAN SOUTHERN OBSERVATORY / SCIENCE PHOTO LIBRARY; 228 Classic Image / Alamy Stock Photo; 229 NASA / JPL / STScI / AURA; 230 tl NASA / CXC / SAO / JPL-CALTECH / MPIA, CALAR ALTO, O KRAUSE ET AL / SCIENCE PHOTO LIBRARY, bl NASA / CXC / RUTGERS / K ERIKSEN ET AL / DSS / SCIENCE PHOTO LIBRARY, b A BURROWS / ARIZONA UNIVERSITY / SCIENCE PHOTO LIBRARY; 231 A BURROWS / ARIZONA UNIVERSITY / SCIENCE PHOTO LIBRARY; 232 HE FLASH CENTER / UNIV. OF CHICAGO / SCIENCE PHOTO LIBRARY; 235 tl World History Archive / Alamy Stock Photo, tr World History Archive / Alamy Stock Photo, bl Gibon Art / Alamy Stock Photo, br World History Archive / Alamy Stock Photo; 236 t Photo 12 / Alamy Stock Photo, b Photo 12 / Alamy Stock Photo; 237 Science History Images / Alamy Stock Photo; 238 t Keystone Press / Alamy Stock Photo, b GL Archive / Alamy Stock Photo; 239 Keystone Press / Alamy Stock Photo; 240 RGONNE NATIONAL LABORATORY / SCIENCE PHOTO LIBRARY; 242 t ESA PLANCK COLLABORATION / SCIENCE PHOTO LIBRARY, b ESA, HFI & LFI consortia (2010); 244 tl ESA - S Corvaja, tr PATRICK DUMAS / LOOK AT SCIENCES / SCIENCE PHOTO LIBRARY, b ESA, S Corvaja, 2009; 245 PATRICK DUMAS / LOOK AT SCIENCES / SCIENCE PHOTO LIBRARY; 246 PLANCK COLLABORATION / ESA / SCIENCE PHOTO LIBRARY; 248 t ESA / Planck Collaboration, b NASA Goddard; 249 JSC / Reid Wiseman; 250 Hua Zhu / Solent News / Shutterstock